Pollution Control in the United States

Evaluating the System

J. Clarence Davies and Jan Mazurek

Resources for the Future
Washington, DC

Printed in the United States of America

Published by Resources for the Future
1616 P Street, NW, Washington, DC 20036–1400

Library of Congress Cataloging-in-Publication Data

Davies, J. Clarence, 1937–
    Pollution control in the United States / by J. Clarence Davies and Jan Mazurek.
        p.   cm.
    Includes bibliographical references (p.   ) and index.
    ISBN 0–915707–87–X (cloth) — ISBN 0–915707–88–8 (pbk.)

    1. Pollution—United States—Evaluation. 2. Pollution—Government policy—United States—Evaluation. 3. Pollution—Law and legislation—United States. I. Mazurek, Jan, 1965–   . II. Title.
TD180.D39   1998   97–49246
                                                                         CIP

The paper in this book meets the guidelines for permanence and durability of the Committee on Production Guidelines for Book Longevity of the Council on Library Resources.

This book is the product of the Center for Risk Management at Resources for the Future, J. Clarence Davies, director. It was typeset in Palatino by Betsy Kulamer; its illustrations were prepared by Bevi Chagnon, Pubcom; and its cover was designed by Diane Kelly, Kelly Design.

RESOURCES FOR THE FUTURE

Directors

About
Resources for the Future

Resources for the Future is an independent nonprofit organization engaged in research and public education with issues concerning natural resources and the environment. Established in 1952, RFF provides knowledge that will help people to make better decisions about the conservation and use of such resources and the preservation of environmental quality.

RFF has pioneered the extension and sharpening of methods of economic analysis to meet the special needs of the fields of natural resources and the environment. Its scholars analyze issues involving forests, water, energy, minerals, transportation, sustainable development, and air pollution. They also examine, from the perspectives of economics and other disciplines, such topics as government regulation, risk, ecosystems and biodiversity, climate, Superfund, technology, and outer space.

Through the work of its scholars, RFF provides independent analysis to decisionmakers and the public. It publishes the findings of their research as books and in other formats, and communicates their work through conferences, seminars, workshops, and briefings. In serving as a source of new ideas and as an honest broker on matters of policy and governance, RFF is committed to elevating the public debate about natural resources and the environment.

Contents

Part III: Overview

Foreword

Mark Twain once observed that everyone complains about the weather, but no one does anything about it. Analogously, whenever experts are polled about how the United States might better manage its environment—and its environmental protection programs—they almost always include on their list the need for careful *ex post* analyses of existing programs. How, they ask, can we design successful programs for the future if we fail to take the time to see what has and has not worked in the past? Yet seldom does one see even preliminary versions of such evaluations.

Terry Davies and Jan Mazurek have done something about this shortfall in evaluation—and in a grand way. *Pollution Control in the United States: Evaluating the System* is an extraordinarily ambitious book. Indeed, it attempts to identify the successes and failures of the whole corpus of environmental laws and regulations that have been put in place over the last twenty-seven years in the United States. They appraise statutes and subsequent regulations aimed at protecting the environment through control of air and water pollution, solid and hazardous wastes, pesticides and toxic substances, and a variety of other possible threats.

The factors at work against environmental program evaluation are diverse. One would, for instance, expect legislators to have strong interest in seeing whether the programs they voted into place are functioning as intended. However, much more credit attaches to creating new laws and programs than to evaluating the performance of existing ones. Apart from an occasional investigation by the General Accounting Office, the legislative branch of government has shied away from careful *ex post* reviews.

At least two things work against careful program evaluation by the Environmental Protection Agency itself. First, Congress is constantly passing new laws or amending existing ones, thereby adding to the already long list of things the agency is required to do. The agency's ever-expanding agenda leaves little time, and virtually no money, for careful retrospective analysis of programs that have been in place for some time.

Second, there is always the risk that such an analysis will put the agency in a bad light. Suppose, for instance, that careful evaluation shows that an air or water pollution program has been both quite expensive and quite ineffective at improving air or water quality. Embarrassment would attend EPA and, possibly, those who had championed the laws giving rise to the program. Why run the risk of such embarrassment, some argue, when many new initiatives require analytical attention?

Fortunately, the Andrew W. Mellon Foundation shares neither pride of authorship of this nation's environmental laws nor fear of exposing whatever failures may exist. For that reason, the foundation asked Terry Davies of Resources for the Future in 1994 whether he would conduct a thorough assessment of the pollution control policies of the United States. Davies, who was "present at the creation" as one of those who helped create EPA via a reorganization of the executive branch of the federal government in 1970, was willing and able to take on such an assignment. He quickly teamed up with Jan Mazurek, his then-colleague in RFF's Center for Risk Management, to begin the work. We at RFF are immeasurably grateful to the Andrew W. Mellon Foundation and the Smith Richardson Foundation for their generous support of this project.

Pollution Control in the United States is the product of Terry Davies's and Jan Mazurek's considerable efforts. It painstakingly describes the laws and major regulations that are the backbone of pollution control programs in the United States; the process through which regulations are developed and analyzed within EPA and other sectors of the executive branch; and the critical role played in this process by the states.

The book also identifies a set of criteria against which the country's various pollution control programs can be measured. Importantly, these criteria could also be used to evaluate other government programs, whether in housing, crime prevention, education, or agriculture. Davies and Mazurek measure our air and water pollution control programs, as well as those designed to reduce risks from other environmental pollutants, against the qualitative "benchmarks" they have identified.

The results of this herculean effort will be of great value to those in the executive and legislative branches of the federal government, to those in the environmental and business communities, to academics interested in environmental policy or program evaluation, to reporters who cover environmental issues, and to ordinary citizens as well. In fact, because it is clearly written, carefully documented, and quite frank, the book's greatest value may be to those in the last, comprehensive group. After all, it is as ordinary citizens that we pay for our environmental protection efforts (through higher product prices, reduced shareholder earnings, or diminished job opportunities) and enjoy the considerable benefits associated with clean air and water, uncontaminated food and soil, and safe consumer products. It is appropriate, then, that this book is so accessible to "real people."

Paul R. Portney
President
Resources for the Future

Preface

Even more than most books, this volume represents the collective efforts of a large number of people. The initial idea for such a comprehensive study came from the Andrew W. Mellon Foundation, and we are grateful to Bill Robertson of that foundation for his support and guidance. We never would have undertaken this effort without the Mellon Foundation's initiative. The other key funder who made the project possible was the Smith Richardson Foundation, and we are indebted to Mark Steinmeyer and the others at Smith Richardson.

The research and writing of the report was a collective effort. The research and initial drafting of the efficiency chapter were done by Dick Morgenstern and Liz Farber, and Dick was continuously helpful with wise counsel and advice. Michael McGovern wrote the public involvement and nonintrusiveness sections of the social values chapter and Bob Hersh wrote the environmental justice section. The initial drafts of the chapter comparing the United States to other nations were done by Elise Annunziata. The futures chapter was initially drafted by Kevin Milliman. Other major research support came from Mark Powell, Tom Votta, Todd Straus, Kieran McCarthy, and Nicole Darnall. Skillful editing was provided by Rich Getrich and Eric Wurzbacher.

All or parts of the manuscript were reviewed by a number of people who provided helpful comments. We are grateful to Paul Portney, Chris Kelaher, Allen Kneese, Michael Kraft, Mike Newman, Kate Probst, Vic Kimm, Marilyn Voigt, Pete Andrews, David Clarke, Mike Taylor, and Ruth Bell for advice and assistance. John Mankin carefully and patiently typed numerous drafts.

Any praise for this work therefore should be shared among a large number of people. As is usual, customary, and fair, any blame should be directed exclusively at the two authors.

J. Clarence Davies
Jan Mazurek

Pollution Control
in the United States

1
Introduction

This book is both an evaluation of the past and a road map for the future. By identifying the parts of the pollution control regulatory system that have worked and the parts that are "broken," the book provides a guide to where future reforms should be focused. In a period of political history when a lot of bath water is being thrown out, creating the possibility that many babies are also being discarded, we hope that the report will maximize the chances of preserving good policies while encouraging radical change in a system that badly needs changing.

THE NEED FOR EVALUATION

Serious objective evaluation of public programs is a rare activity. As a general rule, organizations that have a stake in a particular program do not want to jeopardize their interests by stepping back to evaluate whether the program is working. Organizations that do not have a stake tend not to have an interest in evaluating the program. Although a few organizations in the federal government have an explicit responsibility for program evaluation (the General Accounting Office is a notable one), even these organizations rarely undertake broad, probing program evaluations.

The dearth of evaluation certainly applies to the pollution control regulatory system. For many years the Environmental Protection Agency (EPA) had a program evaluation division, but it essentially functioned as a management consulting group for the EPA programs, performing almost no real program evaluation. (See NAPA 1995, 168–69.) The division was finally abolished in 1994.

The General Accounting Office has evaluated a number of aspects of the pollution control system, and has done an overall review of EPA activities (U.S. GAO 1988). Also, the Environmental Working Group (1994) has issued annual reviews of EPA's work. While a number of individual scholars have examined EPA and pollution control policies, evaluation has not been a major purpose of most of these studies. (There are some notable exceptions. See Landy, Roberts, and Thomas 1994; Hahn 1994; Vig and Kraft 1997; and Knapp and Kim forthcoming.)

The present report is the most comprehensive evaluation of the pollution control regulatory system ever conducted. In fact, it is one of the broadest attempts at program evaluation for any program area. (See Rossi and Freeman 1993.) We hope that, in addition to its primary purpose of evaluating pollution control policy, the study will also contribute to the methodology of program evaluation generally.

THE POLLUTION CONTROL REGULATORY SYSTEM

The U.S. pollution control regulatory system is large and complex. It is certainly debatable whether in most senses it can even be called a "system." We consider it to be an interrelated whole because its parts possess a common set of purposes and its constituent groups have a relatively high degree of interaction. We use the term "system" in a general sense, that is, as a set of entities that interact with each other and whose behavior cannot be described or understood apart from the other entities in the system—in von Bertalanffy's words, "complexes of elements standing in interaction" (von Bertalanffy 1968, 33).

At the center of the system is the federal Environmental Protection Agency. Several later chapters will examine the character and functioning of EPA in some detail. EPA is small compared with other major federal agencies. In 1996, it had 17,200 employees; for comparison, the Department of Agriculture had 100,700 employees, and the National Aeronautics and Space Administration 21,100 (U.S. OMB 1987, Table 10-1, 206). EPA is headed by an administrator, who reports to the President.

EPA's function is to implement the pollution control laws enacted by Congress. Its most important function in this respect is to establish national standards that govern how much pollution is allowable. It also promulgates a large number of regulations that prescribe how the standards are to be defined and how compliance is to be established, as well as an enormous number of other requirements for reporting and other processes.

The laws that prescribe the federal pollution control functions are byzantine. They result from more than forty years of accumulated lawmaking by Congress. During that period, Congress has made no effort to simplify or pare down the legislative provisions, and the scope and complexity of environmental law has steadily increased. It is ironic but not surprising that the regulatory reform legislation considered in the 104th

Congress, which many members of Congress supported as a way of reducing regulatory activity, would have added to the volume and complexity of the pollution control legal structure.

State pollution control laws generally are designed to implement the framework prescribed by the federal statutes, although each state contributes its own unique touches. For most environmental programs, state standards are permitted to be more stringent than federal standards but not less stringent. In recent years, a number of states have passed legislation prohibiting state standards from being more stringent than the federal standards.

Most of the permitting, enforcement, and other daily activities necessary to implement both the national and state standards are carried out by the fifty states, each of which has its own pollution control agency or agencies. As is typical in the American intergovernmental system, none of the basic pollution control functions is solely a state or solely a federal responsibility. All the functions are to some degree shared.

An extensive number of nongovernmental groups have some interest in pollution control. All business groups are subject to regulation except for small local enterprises—and even some of them are regulated. The government pollution control agencies have to cope with the large number and diversity of business organizations within their jurisdiction for which they are responsible for having some knowledge. On the other hand, because the number and diversity usually mean that some corporations will gain and others will lose from almost any pollution control program, the business community is often politically divided.

Environmental groups in the United States are as diverse as business groups. There are hundreds of national environmental groups and thousands of local ones. Politically, environmental groups cover a very broad spectrum ranging from radical antigovernment groups to groups led by the aristocratic elite. A growing split has been developing among national environmental groups in recent years, with some of the newer groups such as Greenpeace and the Public Interest Research Group accusing older groups such as the National Wildlife Federation and the Audubon Society of having "sold out" to the government and big business. Some groups focus on pollution generally, while the primary interest of others may be on wilderness or endangered species or dam construction. Groups that are not primarily concerned with the environment, such as the American Lung Association or the Competitive Enterprise Institute, sometimes take an active role in pollution control issues.

The complex legal framework that governs pollution control provides numerous and easy avenues for groups to use litigation as an instrument to further their ends. All major parties in the pollution control system (including the government) frequently resort to the courts, thus making the courts themselves major actors in the system. Judges ultimately make a lot of pollution control policy (see McSpadden 1997).

To summarize, we have defined the boundaries of the pollution control regulatory system by those problems and policies generally within the jurisdiction of the federal EPA. The actors in the system, in addition to EPA, include Congress, the courts, state pollution control agencies, and nongovernmental parties that have an interest in pollution control policies.

To keep the study manageable, we have excluded here three relevant categories of problems or policies. The first limitation is that we have focused on pollution control, not on the environment. A number of environmental policies and issues, such as parks and wilderness, endangered species, natural resource availability, and land use, fall outside our definition of the system we are examining. Such topics have close and important ties to pollution problems, but they involve different laws, agencies, and constituencies, and we have generally excluded them from our consideration. (An exception is the examination of water quantity in the futures chapter, because it is impossible to deal with the future of water quality without examining the future of water quantity.)

A second exclusion involves the broad underlying factors that drive much of the generation of pollution, particularly population growth and technological change. Although we believe that population growth is not only a major problem in itself but a source of many pollution problems, an evaluation of population policy would be quite different from an evaluation of pollution policy. Technological change is a more complex factor because it both creates pollution problems and is a key part of solving such problems. In various parts of this study we deal with technological change because it is so intrinsically related to pollution policy. However, as with population, we are trying to study and evaluate pollution policy, not technology policy, and the two are not the same.

The third category of exclusions is the most troublesome. We believe that the future of pollution control and environmental quality in the United States (the U.S. focus being another limitation of our system definition) will be determined far more by energy policy, agricultural policy, transportation policy, and other sectoral actions than by EPA's regulatory policies. Regulations can affect these other policies and actions, but if one were trying to anticipate the government's impact on future environmental conditions it would be more fruitful to examine the Department of Energy than EPA. We will deal with these interconnections at various points in this study. However, we could not study the pollution control regulatory system *and* study the energy policy system or the agricultural policy system.

How can we evaluate the cumulative effect of all these groups and laws and actions? Several criteria allow us to make sense of the system and pass some judgments on it.

CRITERIA FOR EVALUATION

Two commonly used criteria for judging the pollution control regulatory system are effectiveness (is the system meeting its goals) and efficiency (are system outputs being accomplished with the minimum necessary resources). If the system is not effective, it obviously is deficient. If it is effective but inefficient, it is using valuable societal resources that could be employed for other purposes.

Effectiveness in this context has two dimensions. The first is whether the system is accomplishing the goals and objectives specified in legislation and regulation. These are the criteria of effectiveness the system has set for itself. We use the available monitoring and other data to gauge the extent to which the system has met its own goals.

The second dimension of effectiveness is whether the goals on which the system focuses are the "right" goals. If the overarching goal is to reduce risk or protect the environment, has the system really concentrated its efforts on the highest risks, the greatest environmental threats? It is quite conceivable that the system could accomplish all its objectives and that environmental risk would still have increased because the system was not targeted at the most important risks. We use a comparative risk analysis to examine this issue.

Efficiency also is a multidimensional criterion, involving the relationship of costs to benefits, of costs to reach a given objective (cost-effectiveness), the effect of pollution control costs on international competitiveness, as well as other considerations. We have used both existing literature and original analysis of the available pollution control cost data to measure the system's efficiency.

Other criteria apart from efficiency and effectiveness are important and relevant. One is the future dimension. The system might be performing quite well now but be destined for trouble in the future. Future problems could be created either because of new problems on the environmental agenda (stratospheric ozone depletion was not a perceived problem in 1960) or because economic growth, population growth, or other factors may aggravate existing problems beyond the capability of the existing system to deal with them. An example of the latter would be if automobile vehicle miles traveled increased at such a pace that levels of smog and other auto-related air pollutants significantly worsened. While predicting the future is inherently difficult, we have attempted to look ten to twenty years ahead to identify trends or problems that might cause difficulty for the current control system.

Another relevant criterion is how the United States compares with other developed nations. This comparative judgment of the U.S. system may suggest ideas for improving the system. We use the existing literature, especially the environmental performance review reports of the Organisation for Economic Co-operation and Development, to compare U.S. pollution control programs with those of other countries.

Finally, several values broadly held in American society are important criteria used by citizens in judging pollution control efforts. One is equity, how fairly the costs and benefits of pollution control are distributed. The *environmental justice* movement has made equity considerations the basis of its harsh criticisms of current pollution control. Another value is *nonintrusiveness*, the extent to which control efforts respect people's privacy and do not impinge excessively on their time or freedom. The perceived intrusiveness of the current system is a fundamental reason for its political unpopularity among certain groups. Other widely held values that also need to be considered are *public participation*, the extent to which the system educates and empowers the citizenry, and the degree to which the system encourages *cooperation* in contrast to hostility and conflict. Our report tries to analyze how the pollution control system is doing when judged by these societal values.

Two more general questions arise when employing these evaluative criteria. First is the question of *what* is being evaluated. We noted above a variety of important areas that were excluded from our definition of the pollution control regulatory system. However, even within the system there is the further question of whether we are evaluating the system as a whole or its individual components.

There are several ways to define the components of the system—different programs, different actors, different geographical areas—but programs is probably the most useful concept. The distinction between evaluating the system and evaluating individual programs is important because different programs fare differently when judged by our criteria. For example, we might conclude that the Superfund waste cleanup program is quite inefficient but that the mobile source air pollution control program is efficient. On a more detailed level, we might find that while the mobile source program as a whole is efficient, the current inspection and maintenance program is quite inefficient. The difficulty with our approach is that there are a large number of pollution control programs (more than thirty even by a broad definition), and it is beyond our capability to examine each one in detail. Accordingly, we have examined individual programs selectively, while looking for more general patterns that allow us to draw some general conclusions about the system as a whole.

Evaluation inevitably raises the question "compared with what?" Most program evaluations compare conditions before a social intervention with conditions after the intervention. (See Rossi and Freeman 1993, 34–55.) The large number of diverse interventions encompassed in the pollution control regulatory system makes such an approach difficult. Instead, we have used a set of evaluative criteria each of which is a yardstick independent of alternatives: efficiency, or intrusiveness, can be measured in some absolute sense.

The yardsticks that we have used here represent criteria that analysts might use in evaluating a government program. Whether these are in fact the criteria that people in general would employ (and how they would weight the relative importance of the criteria) is an empirical question that has not, so far as we are aware, been put to the test. In a future research project we hope to examine this question of criteria more closely. For the purpose of this study we identified what we perceive to be the most frequently employed criteria. The reader can judge the extent to which we have succeeded.

STRUCTURE OF THE BOOK

The book has two main parts. The first part looks at the key processes and institutions involved in controlling pollution. It examines federal legislation, administrative decisionmaking (especially at EPA), and the federal-state division of labor. This part of the book provides a context and description for the reader who is not that familiar with the pollution control regulatory system. However, these chapters are evaluative as well as descriptive. *How* the outputs of the system are produced may be as important as *what* the system produces, so it is important to evaluate the process.

The second part of the book evaluates the pollution control system against the evaluative criteria discussed above. There are chapters on reducing pollution levels, targeting the most important problems, efficiency, social values, comparison with other countries, and ability to meet future problems

A final, concluding chapter summarizes the results of our study. Not surprisingly, we find both strengths and weaknesses in the system. It is inefficient, but it is also open to public input. Its priorities are wrong, but it has been effective in dealing with some important problems. It is probably able to deal with the problems of the next couple of decades, but it is woefully lacking in the data necessary to judge how it is doing now. The final chapter also outlines some guiding principles that follow from our analysis. The primary purpose of the study is to analyze and evaluate rather than to formulate recommendations, and detailed recommendations must await a future project. However, we hope that this study provides the requisite groundwork to prepare for the changes that the pollution control system so urgently needs.

REFERENCES

Environmental Working Group. 1994. *Annual Review of the U.S. Environmental Protection Agency.* Washington, D.C.: Environmental Working Group.

Hahn, Robert W. 1994. United States Environmental Policy: Past, Present and Future. *Natural Resources Journal* 34(Spring 1994): 305–48.

Knapp, Gerritt J., and T. John Kim, eds. Forthcoming. *Environmental Program Evaluation: A Primer.* Champaign: University of Illinois Press.

Landy, Marc K., Marc J. Roberts, and Stephen R. Thomas. 1994. *The Environmental Protection Agency: Asking the Wrong Questions.* Rev. ed. New York: Oxford University Press.

McSpadden, Lettie. 1997. Environmental Policy in the Courts. In Norman J. Vig and Michael E. Kraft (eds.), *Environmental Policy in the 1990s.* Washington, D.C.: Congressional Quarterly Press.

NAPA (National Academy of Public Administration). 1995. *Setting Priorities, Getting Results: A New Direction for EPA.* Washington, D.C.: NAPA.

Rossi, Peter H., and Howard E. Freeman. 1993. *Evaluation: A Systematic Approach.* 5th ed. Newbury Park, California: Sage Publications Inc.

U.S. GAO (General Accounting Office). 1988. *Environmental Protection Agency: Protecting Human Health and the Environment through Improved Management,* GAO/RCED-88-101. Washington, D.C.: U.S. GAO.

U.S. OMB (Office of Management and Budget). 1987. The Budget for FY1988. In *Analytical Perspectives.* Washington, D.C.: U.S. GPO.

Vig, Norman J., and Michael E. Kraft, eds. 1997. *Environmental Policy in the 1990s.* 3rd ed. Washington, D.C.: CQ Press.

Von Bertalanffy, Ludwig. 1968. *General System Theory.* New York: George Braziller.

PART I:
Evaluating the Process

2

Federal Legislation

Federal pollution control laws are the bedrock, the driving force, of the country's pollution control system. The internal organization of EPA, its priorities and procedures, and the actions of the states are all driven to a significant extent by the statutory framework. It is hard to imagine any major change in the pollution control system that does not involve a change in federal pollution control laws. This chapter provides an analytic overview of the federal laws and then discusses the role of the courts in the regulatory system.

DESCRIPTION

The federal pollution control laws are so fragmented and unrelated as to defy overall description. There are about nine major laws (see Table 1), although the number could be larger or smaller because the definition of "major" is inherently arbitrary. Added to these are dozens of lesser laws and hundreds of minor ones. Among the laws not included in Table 1 are the Emergency Planning and Community Right-to-Know Act of 1986, the Pollution Prosecution Act of 1990, the Environmental Programs Assistance Act of 1984, the National Environmental Education Act, the Radon Gas and Indoor Air Quality Research Act of 1986, the Oil Pollution Act of 1990, the Rivers and Harbors Act of 1899, the Ocean Dumping Act, and the Environmental Research and Development Demonstration Act.

Congress considers hundreds of environmental bills and sometimes enacts as many as twenty or thirty each session. Most new laws are relatively minor. Bills that became law in the 103rd Congress included the Federal Employees Clean Air Incentives Act (107 Stat. 1995), the Indian Lands Open Dump Cleanup Act of 1994 (108 Stat. 4164), and the Ocean Pollution Reduction Act (108 Stat. 4396). The last is less than two pages in length.

Of the major statutes, three (Clean Air, Clean Water, Safe Drinking Water) are based on the medium in which pollution occurs; one (Resource Conservation and Recovery) is focused primarily on a

11

Table 1. Major Federal Pollution Control Laws.

Name	Date enacted	History
Federal Insecticide, Fungicide, and Rodenticide Act	1947	Originally Insecticide Act of 1910; Major rewrite in 1972 (Federal Environmental Pesticides Control Act); amended in 1964, 1978, 1988, 1991, 1996
Water Pollution Control Act	1948	Some roots in 1899 Refuse Act; became Clean Water Act in 1972; major amendments in 1956, 1965, 1977, 1987
Air Pollution Control Act	1955	Became Clean Air Act in 1963; major amendments in 1967, 1970, 1977, 1990
National Environmental Policy Act	1970	Amended in 1975, 1977
Safe Drinking Water Act	1974	Original provisions in 1944 Public Health Services Act; amended in 1976, 1977, 1986, 1996
Resource Conservation and Recovery Act	1976	Solid Waste Disposal Act of 1965; major amendments in 1984, 1986; minor amendments in 1996
Toxic Substances Control Act	1976	Amended in 1986, 1988, 1992
Comprehensive Environmental Response, Compensation, and Liability Act of 1980	1980	Major amendments in 1986; various minor amendments
Pollution Prevention Act	1990	

medium—land—but deals with other matters as well; one (Insecticide, Fungicide, and Rodenticide) deals with a particular set of products; one (Toxic Substances) deals with chemicals in general; one (Environmental Response, Compensation, and Liability) deals with accidents, spills, dumpsites, and liability; and two (National Environmental Policy and Pollution Prevention) deal with general policy. The inconsistency of the bases of these acts is aggravated by numerous minor statutes, each of which has its own particular perspective on pollution control.

The major pollution control laws are quite detailed and have been getting more so. The full Clean Air Act, prior to the 1970 amendments, was only twenty-two pages long and contained six deadlines, three of them related to studies. The 1970 amendments are thirty-eight pages long and contain some twelve deadlines. The 1990 amendments to the Clean

Air Act are more than 300 pages long and add 162 statutory deadlines to EPA's workload (U.S. EPA 1994). The increasing detail reduces EPA's own flexibility and the flexibility it can allow to regulated entities.

The statutory detail reflects a basic congressional mistrust of EPA. The mistrust is balanced: pro-environment members fear that the agency will not be ardent enough in defending the environment; anti-environment members fear that it will be too ardent. Pro-environment members therefore write detailed marching orders into law, while anti-environment members try to assure that the agency will not have enough resources to fully implement the laws and that courts will have authority to second-guess any agency decisions. EPA is thus the focus of an odd and intricate checks-and-balances system (Lazarus 1991).

HISTORY

The large tree of environmental law has several different roots. One is the common law of nuisance. Another is public health law. A third is attempts to protect navigable waters for navigation, recreation, and other uses. These different roots explain much of why the current system is based on a medium-by-medium, program-by-program approach.

The mythological account of environmental law springing full-blown from the U.S. Congress following Earth Day in 1970 is a distortion. A major drive for a federal water pollution control law occurred prior to World War II, and Franklin Roosevelt vetoed a bill, passed by both houses of Congress, that would have given strong water pollution control authority to the federal government. The effort was renewed after the war, and the Federal Water Pollution Control Act became law in 1948. The Clean Air Act dates from 1955, the Solid Waste Disposal Act from 1965. By 1969 the federal government, usually in partnership with the states, was involved in regulating most major forms of pollution (Davies and Davies 1975, Ch. 2).

The federal-state balance did shift significantly in 1970 as public opinion pushed Richard Nixon and Congress to assert a stronger federal role. A major rewrite of the air act was passed in 1970, and a similar overhaul of the water act passed in 1972. Under the leadership of the Council on Environmental Quality (created in 1970), the Nixon administration proposed a broad legislative program that included major proposals dealing with pesticides, noise, toxic substances, and land use. All but the land use bill eventually became law.

Despite many earlier initiatives and precedents, much of the agenda of environmental problems to be addressed was established in the early 1970s. Some important new topics emerged—hazardous waste sites, pollution prevention, stratospheric ozone depletion, climate change, chemical plant accidents, medical wastes—but the basic topics and framework of legislative debate did not change substantially for twenty-five years. Every few years individual laws would be reviewed and various "strengthening" amendments considered. The process was one of ratcheting ever more stringent provisions with more federal control and more detailed requirements for both polluters and EPA. Among the few exceptions to this general trend was the Noise Control Act (P.L. 92-574; 40 CFR 209), which had not had much political support when first enacted in 1972. The lack of constituency support grew more obvious as EPA tried to regulate noise from various products and Congress steadily reduced appropriations to implement it. Eventually, with very little controversy, the act became a dead letter, with no attempt by EPA to implement its provisions.

The tendency to make environmental laws ever more stringent came to an abrupt halt with the 1994 congressional elections. With Republicans gaining control of both houses, the question became whether the laws would be weakened in fundamental ways, made more effective through bipartisan cooperation, or left unchanged because of governmental gridlock. At the conclusion of the 104th Congress in 1996, it actually appeared that the laws might be made more effective. Both the pesticide laws and the Safe Drinking Water Act were amended in constructive ways. The many proposals to fundamentally weaken the environmental statutes failed to pass both houses. (For more details, see Vig and Kraft 1997.)

The future of environmental law depends largely but not entirely on political trends. The strong latent feeling held by the American people that environmental protection is an essential governmental function can be activated by events in the natural world, at least if the press devotes attention to such events. While no one-to-one correlation exists between the severity of environmental disturbances and the amount of attention they receive, one of the political advantages of many environmental issues is that they lend themselves to press imagery. Any new environmental disaster is likely to affect the political system.

International actions are increasingly likely to affect the directions and policies of EPA. The Montreal Protocol on Substances that Deplete the Ozone Layer has established U.S. policy on chlorofluorocarbons (CFCs) and other chemicals that reduce stratospheric ozone. The North American Free Trade Agreement with Canada and Mexico contains a variety of environmental provisions and was coupled with a North American Agreement on Environmental Cooperation that, among many provisions, established a permanent trinational Commission for Environmental Cooperation (U.S. Government 1993; CEC 1996).

COMMAND AND CONTROL

The current body of pollution control law, both federal and state, has been characterized as "command and control." This vivid phrase has several different meanings. One is that the pollution control system relies on laws, regulations, and penalties—that it is based on potential coercion rather than voluntary goodwill and on penalties rather than on positive incentives.

Although these generalizations are true, they also are somewhat over-simplified. The pollution laws, like any social rules, require a high degree of voluntary compliance for their success. The resources necessary to actually inspect and prosecute all, or even most, pollution sources will never be available. The ultimate penalties that EPA can invoke against recalcitrant states—takeover of the state program or withholding of federal grant monies—can be invoked only very rarely. Thus, while the system is coercive in form, it relies on a large degree of voluntary compliance.

The penalties versus positive incentives contrast also is not as stark as it appears. Command-and-control is often contrasted with market approaches, but most of the market approaches that have been used in the United States operate within the standard command-and-control framework. For example, the sulfur dioxide allowance system under the 1990 Clean Air Act Amendments is inseparable from the standards and penalties provisions of the act. Market mechanisms that are not tied to the regulatory system do exist (for example, deposit-refunds on bottles or increases in the price of gasoline), but they are difficult to formulate and often are stoutly resisted by the entities to which they would apply.

These characteristics do tell us something about the strengths and weaknesses of the current system. Voluntary compliance has significantly reduced pollution below what it would otherwise be (see Chapter 5), but noncompliance nevertheless is a serious problem. In 1994, 27 percent of major water pollution sources were in significant noncompliance with water quality standards (U.S. GAO 1996, 5). Two-thirds of these sources were municipal wastewater treatment plants, while one-third were private companies (U.S. EPA 1997a). EPA has raised questions about the adequacy of enforcement in several states (see Chapter 4).

The limited use of market mechanisms illustrates the inefficiency of many aspects of the current system. Carlson and others (1997), using conservative assumptions, predict that the sulfur dioxide allowance system will save between $450 million and $2 billion per year in 2010, as compared to what control costs would be without its flexibility and market-like approaches. We will deal with efficiency in more detail in Chapter 7.

Another meaning of command-and-control is that such laws and regulations dictate not only the goals to be reached but also the means that must be used to reach them. Many pollution control laws are technology-based, defining goals by the amount of control that current technologies can achieve. In form, the regulations that implement the technology-based standards do not dictate the means to be used, usually allowing any method to be used as long as it will achieve the specified level of control. However in practice, permit writers, inspectors, and consultants tend to take a dim view of any method for meeting the standard other than the technology on which the standard is based. Pollution sources employing alternative methods do so at considerable risk.

The laws and regulations may limit flexibility in other ways. A study conducted by EPA and the Amoco Oil Company of an Amoco refinery in Yorktown, Virginia provides a classic example. EPA regulations required Amoco to control VOCs (volatile organic compounds) emitted from its waste treatment plant—a minor source—but not from the refinery's barge unloading facilities, the second largest source of VOCs. Controls on the unloading facilities, combined with a few other changes, would have allowed Amoco to meet 97 percent of its VOC reduction requirements at 25 percent of the cost (Amoco 1992, 11).

Technology-based provisions, like most aspects of the pollution control laws, may significantly restrict the flexibility of both the regulators and the regulated. They also encourage end-of-the-pipe controls instead of pollution prevention. The fragmentation of the system has the same effect.

FRAGMENTATION

The most obvious and important characteristic of pollution control legislation is its fragmentation. EPA has no organic statute; its mission is not described in any law. The agency's statutory framework consists of dozens of unrelated laws passed at different times by different committees dealing with different subjects.

The negative consequences of this fragmentation include the following:

* *Many existing pollution problems* require an integrated approach for their solution. EPA is unlikely to be able to successfully control pollution from heavy metals or many organic chemicals by focusing exclusively on one part of the environment at a time because these substances are present in air, water, and land and frequently move from one medium to another. In the Great Lakes, whose pollution problems provide insight into those in many other places, toxic pollution comes primarily from air deposition, sediment, groundwater, and land runoff. Some 80 to 90 percent of the PCBs in the lakes is deposited by the atmosphere. The single-medium focus in these kinds of situations will either shift pollution from one medium to

another, sometimes making the situation worse, or simply fail to understand or detect the problem altogether. Polluters are encouraged to dispose of waste in the least regulated medium. Control officials are encouraged to require disposal of waste in any medium except the one for which they are responsible.

- An integrated approach is far more effective in obtaining compliance and reducing pollution. Barry Rabe (1997) has shown that, based on an examination of the integrated permits issued by New Jersey, the existing medium-based system misses about half of the emissions of a typical facility. A materials accounting, integrated approach identifies these "phantom emissions" and thereby subjects them to permitting and control. (See also Rabe 1995.)

- The existing fragmented system makes it impossible to set rational priorities among different programs or even among different control measures. Setting *priorities* necessarily entails some common scale for making comparisons, and degree of risk is the only such scale except for public opinion. Current priorities, however, are based primarily on historical circumstances, custom, and political visibility, not on degree of risk.

- Many of the pollution *control technologies* now used to meet regulatory requirements do not really control the pollution—they simply shift it around, change its form, or delay its release into the environment. A typical wastewater treatment plant controls only about half the toxic substances that go into it. Of the other half, some go to the land in the form of sludge, some are volatilized and becomes air pollution, and the remainder are discharged back into the water. This is no mere theoretical problem: for several years the Philadelphia municipal water waste treatment plant was the largest single source of air pollution in the Philadelphia metropolitan area (U.S. EPA 1986, 11). Fragmentation also makes it more difficult to deal with nonpoint sources of pollution.

- The best way to deal with the problems of current control technology is to shift to an emphasis on *pollution prevention*. Pollution prevention is inherently multimedia, and the current medium-oriented system is a fundamental barrier to implementing prevention strategies. Prevention for stationary pollution sources usually involves changes in plant process rather than end-of-the-pipe solutions, and such process changes would be much more feasible under an integrated approach.

- Considerable evidence indicates that an integrated approach would not only be more effective in protecting environmental quality but would also reduce the *cost* of pollution control and provide more flexibility for industry. The Electric Power Research Institute has shown that an integrated approach to pollution control applied to a new coal-fired power plant could reduce the capital and operating costs of the plant's pollution control system by 25 percent (APCA 1986, 200).

- The current fragmented system is a major impediment to *identifying new environmental problems*. Under the present system, the physical law of the conservation of matter is replaced by the political law of the protection of narrow jurisdictions. Nobody asked what happened to sulfur oxides and nitrogen oxides that were transported long distances because they were beyond the vicinity of the source and therefore no one's concern. That is why recognition of the acid rain problem took a long time. Nobody asked what happened to CFCs after they were released: the problem was not in anybody's regulatory jurisdiction. Thus recognition of the stratospheric ozone problem was similarly delayed. Major problems in the future are likely to be cross-media problems, and they will go unrecognized for a long time if we try to detect them through programs that focus on only one medium of the environment.

- The fragmentation of the pollution control system results in *excessive litigation and bureaucratic red tape*. The existing pollution control system has become so disjointed and cumbersome that no one can understand or make sense of it. A few law firms may benefit, but the situation undermines both compliance and public support.

The negative consequences of fragmentation are of fundamental importance. The need to move toward a more integrated approach is becoming generally recognized. EPA has conducted pilot efforts at integrated approaches for more than twenty years. Its current "Common Sense Initiative" and "XL" projects are only the latest in a long line of such efforts. A National Academy of Public Administration panel recommended to Congress in 1995 that "the environmental control effort should be integrated. In consultation with Congress, and as part of the process of integrating the environmental statutes, the agency should begin work on a reorganization plan that would break down the internal walls between the agency's major 'media' program offices for air, water, waste, and toxic substances" (NAPA 1995, 4).

Internationally, a clear shift is occurring toward holistic approaches to pollution control. Scandinavian countries always have used an integrated approach, the Dutch and Germans are using holistic approaches, the British have recently shifted away from a medium-based approach, and the European Union in 1996 enacted a directive that obligates all member countries to shift to integrated pollution control (Council Directive 96/61/EC, 24 Sept. 1996; see Hersh 1996). U.S. pollution control institutions, once seen as a model for the world, are beginning to be a cumbersome, inefficient anachronism.

OVERLAPS AND INCONSISTENCIES

A corollary of the fragmented legal structure is that it requires EPA to deal with similar situations in different ways and results in duplicative reporting and other requirements. For example, the criterion for setting national standards is different under each major law. National ambient air quality standards under the Clean Air Act are required "to protect the public health" with "an adequate margin of safety" (CAA 109(b)(1)); costs cannot be a consideration. The basic standards under the Clean Water Act are technology-based; cost is implicitly considered in setting the required technology level. The Toxic Substances Control Act and the Federal Insecticide, Fungicide, and Rodenticide Act both explicitly require balancing of costs and benefits before action can be taken.

A different type of inconsistency exists between the Resources Conservation and Recovery Act (RCRA) and the Comprehensive Environmental Response, Compensation, and Liability Act (CERCLA). Both laws may be applicable to the same wastes, for example those at federal facilities, but RCRA responsibilities often are administered by a state agency whereas CERCLA responsibilities cannot be delegated to states. The result is that two different agencies (EPA and the state agency) regulate the same cleanup of the same wastes, and they may disagree on a wide range of requirements (U.S. GAO 1994).

Although these difficulties are significant, the major problems with environmental legislation are their rigidity and lack of coherence. The laws are complex, unrelated to each other, and lacking in any uniform vision of EPA's mission or environmental problems. This is not surprising given that the system of congressional committees and subcommittees dealing with environmental regulation is complex, the committees do not relate to each other, and there is no coherence to the congressional approach. EPA reports to literally dozens of congressional committees and subcommittees.

DISPARITY BETWEEN RESOURCES AND RESPONSIBILITY

Throughout EPA's history, its backers have argued that the money and personnel given to the agency were grossly insufficient to accomplish the agency's statutory mandates. Some have even argued that this disparity is a deliberate congressional strategy to please the public by passing ambitious legislative requirements while pleasing the regulated interests by ensuring that resources are inadequate to implement the requirements (Lazarus 1991).

William D. Ruckelshaus, who served twice as EPA administrator, stated in 1995: "Any senior EPA official will tell you that the agency has the resources to do not much more than ten percent of the things Congress has charged it to do. In addition, they are not empowered to allocate that ten percent so as to ensure a wise expenditure of the public treasure. The people who run EPA are not so much executives as prisoners of the stringent legislative mandates and court decisions that have been laid down like archeological strata for the past quarter-century" (Ruckelshaus 1995, 4). William Reilly, EPA administrator in the Bush administration, has stated simply that "we have got ... to make some allocation of our resources, given the fact that there's more for us to do than we can do" (U.S. EPA 1995, 36).

The total EPA budget is distorted by the large portion that traditionally has gone for water infrastructure grants to state and local governments. In thirteen of the past twenty-five years these grants have accounted for more than half the EPA budget (See Figure 1). EPA's appropriations in FY1997 totaled $6.8 billion, of which $2.2 billion were water infrastructure grants; an additional $1.5 billion were trust funds, primarily Superfund.

In constant dollars, EPA's operating budget (excluding the water infrastructure grants), doubled between 1970 and 1977, doubled again between 1977 and 1987 (with a particularly sharp increase in FY1987), and then leveled out (see Figure 2). Between FY1992 and FY1995, the EPA operating budget stayed approximately level (which means it declined in constant dollars). It increased from $2.6 billion in FY1995 to $3.1 billion in FY1997. When Republicans took control of Congress in 1995, they slashed the agency's total budget, reducing it in FY1996 by almost 20 percent. Almost the entire cut occurred in infrastructure grants; the operating budget actually increased slightly. Agency personnel levels were reduced slightly in FY1996, but by FY1997, they had resumed their modest upward trend (see Figure 3).

No objective criterion exists to evaluate whether the resources available to EPA have been insufficient, too great, or just right. The agency's self-perception is that the resources are inadequate and that difficult choices must therefore be made about which obligations to fulfill. But the perception of difficult choices is accompanied by a lack of adequate internal management systems to determine whether agency dollars and peo-

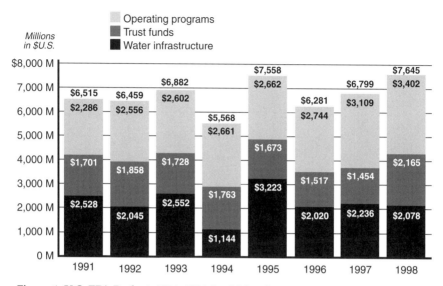

Figure 1. U.S. EPA Budget, 1991–1998, by Major Category.

Source: U.S. EPA 1997b

Note: These figures are per EPA fiscal years.

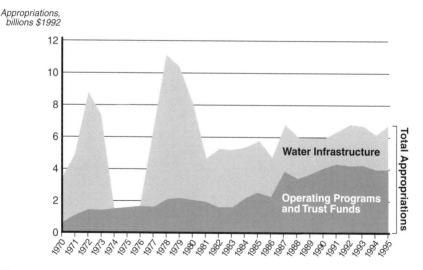

Figure 2. U.S. EPA Budget, 1970–1995 (Constant $1992).

Source: Office of the Comptroller, U.S. EPA, as reported in NAPA 1995, 19.

Note: These figures are per EPA fiscal years.

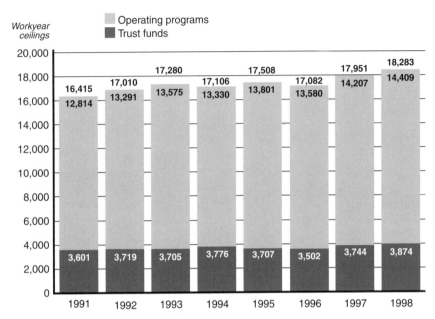

Figure 3. U.S. EPA Personnel, 1991–1998 (Workyear Ceilings).

Source: U.S. EPA 1997b, 8

Note: These figures are per EPA fiscal years.

ple are being used efficiently (NAPA 1994). It does seem certain that the current budget cuts will result in curtailment of some EPA activities.

Increasingly, congressional appropriations bills for EPA have been used both to earmark funds and to promulgate policies. In December 1995, President Clinton vetoed EPA's FY1996 appropriations bill largely because of seventeen separate "riders," substantive provisions added to the bill to control EPA policy (BNA 1996, 1,888). Increasingly, the reports of congressional appropriations committees have contained long lists earmarking specific dollar amounts for specific projects and organizations. For example, the conference report on the FY1997 EPA science budget contained a list of sixteen designated projects; these included $2.5 million for the American Water Works Association Research Foundation, $750,000 for continuation of the Resource and Agriculture Policy Systems Program at Iowa State University, and $1 million for a study of the salinity of the Salton Sea by the University of Redlands (U.S. Congress 1996, H10749). These types of provisions have reduced the agency's ability to set priorities and have resulted in some ill-considered policies.

COURT REVIEW OF EPA DECISIONS

Judges are major decisionmakers in the pollution control regulatory system. Almost all major regulatory decisions are litigated, and EPA's positions are frequently modified or overturned. The most frequently cited figure is that 80 percent of EPA's regulations are challenged in court (Bryner 1994). (According to Michael Kraft, in a personal communication, some consensus has emerged that the 80 percent figure is overstated. Evidence for a lower figure is in Coglianece 1994. We obtained figures from the Department of Justice in an attempt to independently ascertain the correct figure, but the data were not adequate to do so.)

Federal courts have gone through several phases with respect to how much deference they will give to the views of the regulating agency. Glicksman and Schroeder (1991, 249) state: "The stance of the federal courts toward the Environmental Protection Agency has changed substantially. ... An early mix of enthusiasm for the project of environmental protection, respect for the public policy decisions of the Congress, and a rhetoric of close scrutiny of EPA's decisionmaking processes has given way to neutrality toward environmental values, skepticism about whether environmental legislation expresses coherent public policy, and a rhetoric of deference toward EPA's decisions." They also note the relationship between changed judicial attitudes and legislation: "less aggressive judicial review by the courts stimulates Congress to write ever more specific statutes, as it tries to be so specific in expressing its will that even a deferential court will be able to help police agency compliance" (Glicksman and Schroeder 1991, 251).

The trend toward deference to EPA has changed in the 1990s with the politically conservative appointments to the federal courts made by presidents Reagan and Bush and the general shift to conservatism in the political arena. Lettie McSpadden (1997, 184) predicts that "through the remainder of the 1990s it seems likely that most federal courts will scrutinize any regulation of economic behavior whether at the state or federal level. This is likely to take three forms: an increased use of cost-benefit analysis, continued favorable reception to arguments about taking property without due process, and reduced standing for public interest organizations."

A prime example of the extent to which courts may second-guess EPA is the *Corrosion-Proof Fittings* case (947 F.2d 1201 (5th Cir. 1991)), which overturned the ban on various uses of asbestos that EPA had issued under the Toxic Substances Control Act (TSCA). In the case, the Fifth Circuit Court of Appeals, deciding that EPA had acted unreasonably, invalidated most of the asbestos rule. The reasoning of the court is instructive and raises a number of key issues involving the three federal branches.

In the *Corrosion-Proof* case, the court began by citing the difficulty of meeting the substantial evidence test. It then stated that it was invalidating the regulation because EPA had failed to consider all the necessary evidence and failed to promulgate the "least burdensome" reasonable regulation, a TSCA requirement. The opinion did not clearly differentiate between these two categories, but under the failure to consider evidence, it found that:

- EPA failed to obtain public comments on its method for calculating exposure data
- EPA did not consider the toxicity of substitutes for asbestos, even though commentators on the regulation produced evidence showing the toxicity of some substitute products
- EPA failed to consider adequately the costs of the regulation.

The last point led into the court's criticism regarding the least burdensome regulatory alternative. The opinion criticized EPA for failing to calculate the costs and benefits of the regulatory alternatives to banning asbestos. EPA had analyzed five different ban options but not any options for action other than a ban. The court made clear that it considered a cost of $30–40 million per life saved, the calculated cost of some parts of the regulation, to be unreasonable. It observed that EPA should have discounted the benefits of the regulation because it discounted the costs, and that the agency should have calculated the benefits for more than thirteen years into the future because it relied heavily on future unquantified benefits beyond that period.

The *Corrosion-Proof* ruling raises several important questions about the strengths and weaknesses of courts in dealing with environmental regulation. Do courts have the political legitimacy to determine what is a "reasonable" value to place on a human life? Do they have the technical expertise to decide the number of years that should be covered by a benefit-cost analysis? Do they have the administrative knowledge to decide which alternative regulatory approaches should be analyzed? Opinions will differ on the answers to these questions, but they are the kinds of issues that must be raised when considering the role of courts in pollution control regulation.

REFERENCES

Amoco. 1992. *Amoco–U.S. EPA Pollution Prevention Project, Yorktown, Virginia.* Executive Summary. Revised May 1992.

APCA (Air Pollution Control Association). 1986. *Integrated Environmental Controls for Fossil-Fuel Power Plants.* Pittsburgh, Pennsylvania: APCA.

BNA (Bureau of National Affairs). 1996. *BNA Environmental Reporter* 26(38, February 2).

Bryner, Gary C. 1994. Review of Critique of EPA and Environmental Regulation. In U.S. Office of Technology Assessment, *Rethinking Environmental Regulation.* Washington, D.C.: U.S. OTA.

Carlson, Curtis, Dallas Burtraw, Maureen Cropper, and Karen Palmer. 1997. *SO_2 Control by Electric Utilities: What Are the Gains from Trade?* Discussion Paper. Washington, D.C.: Resources for the Future.

CEC (Commission for Environmental Cooperation). 1996. *Annual Report.* Montreal, Canada: CEC.

Coglianece, Cary. 1994. *Challenging the Rules: Litigation and Bargaining in the Administrative Process.* PhD dissertation, University of Michigan.

Davies, J. Clarence, and Barbara S. Davies. 1975. *The Politics of Pollution.* Second edition, Indianapolis: Pegasus.

Glicksman, Robert, and Christopher H. Schroeder. 1991. EPA and the Courts: Twenty Years of Law and Politics. *Law and Contemporary Problems* 54(4): 249–310.

Hersh, Robert. 1996. *A Review of Integrated Pollution Control Efforts in Selected Countries.* Discussion Paper 97-15. Washington, D.C.: Resources for the Future.

Lazarus, Richard J. 1991. The Tragedy of Distrust in the Implementation of Federal Environmental Law. *Law and Contemporary Problems* 54(4): 311–74.

McSpadden, Lettie. 1997. Environmental Policy in the Courts. In Norman J. Vig and Michael E. Kraft (eds.), *Environmental Policy in the 1990s.* Washington, D.C.: Congressional Quarterly Press.

NAPA (National Academy of Public Administration). 1994. *Getting the Job Done.* Washington, D.C.: NAPA.

———. 1995. *Setting Priorities, Getting Results: A New Direction for EPA.* Washington, D.C.: NAPA.

Rabe, Barry G. 1995. *Permitting as a Tool for Integration and Prevention: The Case of New Jersey.* Unpublished paper in authors' files.

———. 1997. Integrated Environmental Permitting: Experience and Innovation at the State Level. *State and Local Government Review.* 27(3): 209–20.

Ruckelshaus, William D. 1995. Stopping the Pendulum! *Environmental Law Forum* 12(6): 25–29.

U.S. Congress. 1996. *Conference Report on H.R. 3666,* September 20, 1996.

U.S. EPA (Environmental Protection Agency). 1986. *Final Report of the Philadelphia Integrated Environmental Management Project.* Washington, D.C.: U.S. EPA.

———. 1994. *Implementation Strategy for the CAA Amendments of 1990.* Update, November. EPA 410-K-94-002. Washington, D.C.: U.S. EPA.

———. 1995. Oral History Interview-4, William K. Reilly, EPA 202-K-95-002. Washington, D.C.: U.S. EPA.

————. 1997a. Data in response to FOIA request from RFF.

————. 1997b. *Summary of the 1998 Budget*. January. EPA 205-5-97-001. Washington, D.C.: U.S. EPA.

U.S. GAO (General Accounting Office). 1994. *Nuclear Cleanup: Difficulties in Coordinating Activities Under Two Environmental Laws*, GAO/RCED-95-66. Washington, D.C.: U.S. GAO.

————. 1996. *Water Pollution: Many Violations Have Not Received Appropriate Enforcement Attention*, GAO/RCED-96-23. Washington, D.C.: U.S. GAO.

U.S. Government. 1993. *The NAFTA Report on Environmental Issues*. November 1993 (no specific agency attribution).

Vig, Norman J., and Michael E. Kraft, eds. 1997. *Environmental Policy in the 1990s*. Washington, D.C.: Congressional Quarterly Press.

3

Administrative Decisionmaking

Much of the recent criticism of the pollution control regulatory system has focused on the decisionmaking process within EPA. Regulatory reform legislation considered by Congress in 1995 and 1996 primarily tried to change how EPA makes regulatory decisions by forcing the agency to do formal risk assessments and more elaborate economic analysis, requiring agency decisions to give greater weight to economic factors, and mandating greater use of peer review and public comments.

Empirical data are scarce on how much effect such changes would have on EPA. The agency makes a large number of decisions of many kinds, and the context of the decisions has changed frequently because of shifts in political climate, changed personalities, and new understanding or interpretation of problems. The result is that views about how EPA currently makes decisions diverge widely, even among observers who would be expected to have similar perspectives. Some think that the agency frequently ignores or distorts relevant scientific information, while others think that it does a good job on considering scientific data. Some think that economic factors are always an important component of decisions, while others believe that economic considerations are usually an afterthought or a rationalization of decisions made on other grounds. Some are impressed with the orderliness and rationality of the decisionmaking process, while others characterize the process as hasty, ad hoc, and largely driven by short-term political considerations.

SCOPE AND LIMITATIONS

By almost any definition, EPA makes a lot of decisions. The new chemicals program under the authority of the Toxic Substances Control Act (TSCA) annually reviews more than 2,000 notifications of new chemicals (Goldman 1994, 12). For each one it must decide whether to seek more information or take some other kind of action. The EPA Pesticides Office in FY1996 registered 24 new active ingredients, registered an additional 701 new products, approved 3,258 amendments to existing pesticide registrations, and made 1,411 other formal registration-related decisions (U.S. EPA 1996a, 30). Thus the agency made more than 20 decisions each working day just on pesticide registrations.

No authoritative figures are available on the total number of regulatory decisions made by EPA each year. The U.S. Office of Management and Budget issues an annual regulatory agenda (U.S. OMB 1997) that shows EPA proposing about 100 new regulations each year, but these are only significant regulations and represent but the tip of a large iceberg.

Environmental statutes are the driving force behind these regulatory decisions. By definition, regulations are simply efforts to apply, clarify, define, and describe the wording of enacted laws. The regulatory agenda is largely determined by laws, and the weight given to various factors in making a decision also is significantly determined by the relevant law.

EPA's decisionmaking discretion varies widely depending on the specific part of the particular law that provides the legal basis for the regulation. Two examples from the Clean Air Act illustrate the range. Some parts of this law specify in great detail what the standard should be. For example, emissions standards for hazardous air pollutants from existing sources must be at least as stringent as "the average emission limitation achieved by the best performing 12 percent" of sources within a given source category (CAA sec. 112(d)(3)(A)). The agency has some flexibility in defining the source category and the way in which the emissions limitation will be measured, but basically the law specifies how EPA is to arrive at the emissions standard. (See, for another good example, sec. 243, Standards for Light-Duty Clean Fuel Vehicles.) In contrast, another part of the Clean Air Act states that "the Administrator shall also examine other categories of sources contributing to nonattainment of the PM-10 standard, and determine whether additional guidance on reasonably available control measures and best available control measures is needed..." (sec. 190). The latter type of wording gives the agency a good deal of discretion. However, as discussed in Chapter 2, the environmental statutes have grown ever more prescriptive and detailed and have granted less and less discretion to EPA.

The internal agency process for making decisions depends on the decision and the decisionmaker. The great majority of regulatory decisions are made at the assistant administrator level or below. The major regulatory decisions most frequently have been made by the deputy administrator presiding over a meeting of the relevant assistant administrators and their staff. An average of fifteen to twenty regulatory decisions a year are made in this way. A small handful of regulatory decisions, perhaps four or five a year, are made by the administrator.

The regulatory process can take a long time because of the complexity and amount of analysis that must be done and because of the need to negotiate the content of a regulation both within EPA and with interested outside parties. EPA often takes several years to formulate a major regulation, and there are examples of more than a decade passing between initiation of work on a regulation and its final promulgation. For example, work on a rule to phase out uses of asbestos began in 1979 (U.S. EPA 1979). The final rule, under the Toxic Substances Control Act, was issued in 1989 (U.S. EPA 1989). Even after the agency promulgates the final regulation, the issue is not settled. Many major regulations are litigated (see Chapter 2), a process that can take years. The asbestos rule was struck down by a 1991 court decision (*Corrosion-Proof Fittings, et al. v. EPA*, U.S. Court of Appeals, Fifth Circuit, No. 89-4596, October 18, 1991).

USE OF SCIENTIFIC INFORMATION

EPA is a regulatory agency dominated by a legalistic culture that generally looks for engineering-based solutions to meet statutory obligations. Typically, the agency's norms, staffing patterns, and incentives subordinate science and scientists within the organization, and economic and political considerations sometimes overwhelm science in regulatory decisionmaking. Research and development accounts for less than 10 percent of EPA's total budget. Furthermore, the core role of the agency's Office of Research and Development, which houses less than 30 percent of EPA's scientists and engineers (the others are in the regulatory programs), has never been resolved.

EPA administrators and legislators are typically attorneys who lack formal scientific training and thus may lack an understanding of what science can and cannot contribute to environmental decisionmaking. To the extent that they rely on science, it is often to defend, attack, or negotiate policy positions.

In part, EPA's lack of scientific orientation is due to large scientific uncertainties that invite the agency and Congress to base their decisions on economic, political, administrative, or technological criteria. But to some degree scientific uncertainty is self-fulfilling, as the inability of science to immediately provide useful answers undermines arguments for allocating to science the time and resources necessary to get the answers. Even in cases where EPA's political appointees demand rigorous and balanced scientific analysis from their staff, the generally limited tenure of the policymakers often makes them unable to take advantage of original data created in response to their perceived informational needs. The short time-horizon of political appointees inside and outside EPA thus helps to explain the chronic failure of strategic planning for environmental regulatory science. In addition, Congress, the courts, and the public have often placed time constraints and additional burdens on EPA that prevent the agency from conducting more thorough scientific analyses in support of its primary regulatory mission.

Legislative provisions can subordinate or promote science in environmental regulatory decisionmaking. Provisions requiring technology-based standards, for example, ensure that engineering and affordability criteria dominate regulatory decisions. In other cases, legislative premises distort the use of science. Statutes (for example, the Clean Air Act) may mistakenly presume that an ambient pollutant concentration always exists below which no adverse health effects will occur. Consequently, standards developed under such statutes would appear to be constrained merely by our technical capabilities to detect biological changes at ever-lower concentrations. Statutory prohibitions of considering costs in setting environmental standards encourage dishonest, pseudoscientific debates that are really about policy choices (that is, who will we protect, and from what).

Factors that appear most critical in facilitating EPA's use of science include the demand for science by EPA's political leadership, peer review of the scientific basis for regulatory decisions, and more sophisticated external scrutiny of the agency's use of science by Congress, the courts, industry, and the public. Adequate time and resources to develop and analyze science are also important. Over time, the use of science at EPA has become more institutionalized, decentralized, consistent (through guidelines), rigorous, and comprehensive (through consideration of multiple pollutants and multimedia exposures).

RISK ASSESSMENT AND COMPARATIVE RISK

EPA pioneered the use of risk assessment to facilitate decisionmaking, and the agency continues to employ it in a variety of contexts. In 1993, according to an agency analysis, EPA completed 7,595 risk assessments. More than 6,000 were quick screening analyses, each of which required no more than two days of work and most of which required only minutes. In contrast, 249 of the assessments were major projects that required more than four person-weeks (NAPA 1995, 37).

The quick screening risk assessments are mostly used in Toxic Substances Control Act (TSCA) programs. Several thousand premanufacturing notifications of new chemicals are screened each year to ascertain whether EPA should require more information from the manufacturer or should prescribe restrictions on the use or handling of the chemical. The TSCA program also screens several thousand existing chemicals each year.

In-depth risk assessments tend to be of two types. One type is used to evaluate site-specific situations, typically at Superfund waste sites. The risk assessment is a component in deciding what remedial measures should be taken at the site, although it is usually not the most important factor (Katherine N. Probst, personal communication). The second type of risk assessment is used in making decisions about nationally applicable regulations. To the extent that such regulatory decisions employ a benefit-cost framework, the benefit is the amount of risk that the regulation will reduce, an amount determined by risk assessment.

A number of risk experts, mostly those employed by regulated industries, have charged that EPA's methods for conducting risk assessments exaggerate the degree of risk. For example, the Hazardous Waste Cleanup Project, an industry coalition, concluded that EPA "has injected a number of assumptions and policy decisions into the scientific exercise of risk assessment, and thus systematically exaggerated the risk posed by contaminants at hazardous waste sites. In fact, EPA commonly overstates the risk at Superfund sites by a factor of 100, 1000, or more" (HWCP 1993, iii). Whether and to what extent this charge is true is the subject of some debate; however, some nonscientific assumptions are an integral part of any risk assessment (McCray 1983), and any statement that refers to risk assessment as a purely scientific process is likely to be a statement crafted for political purposes.

Comparative risk assessment may involve more than just comparing the results of individual risk assessments. Of the several different types of comparative risk assessment (see Davies 1995), the type most relevant for EPA decisionmaking is broad, programmatic comparative risk assessment, a comparison of the risk of broad problems or programs (see Chapter 6). Programmatic comparative risk assessment utilizes data from individual risk assessments but relies heavily on expert judgment and social values. Although EPA has been in the forefront of developing this type of risk comparison, the agency has not made much use of it in setting its own priorities. It is logical to take into account the relative riskiness of different problems when allocating budget dollars; but, while the EPA budget process has varied from year to year, it has never seriously utilized comparative risk data (NAPA 1995, 140–63). In 1996, EPA established a new office that combined the budget and planning functions, so risk may play a larger role in budgeting in the future.

ECONOMIC ANALYSIS

EPA is more advanced than other comparable agencies in performing economic analysis, although the agency can be characterized as ambivalent about the use of economic data and analysis. On the one hand, all agency decisionmakers recognize that cost is the major nonenvironmental factor driving decisions. On the other hand, pro-environment individuals and groups tend to regard cost considerations as a device used to undermine stringent environmental regulations, and many regard benefit-cost analysis with extreme suspicion.

Almost since its creation EPA has been subject to executive branch requirements, usually promulgated by executive order, that the agency do economic analyses for major regulations. (See, for example, Carter's Executive Order 12044, 3 CFR 152 (1978); Reagan's Executive Order 12291, 46 Fed. Reg. 13193 (1981); and Clinton's Executive Order 12866, 58 Fed. Reg. 51735 (1993).) These economic analyses are contained in what are called "regulatory impact analyses" (RIAs). The RIAs have varied widely in quality and impact. Sometimes they are not done until after the decision has been made. (See Morgenstern 1997.)

The fact that agency decisions have been subjected to examination on economic grounds has resulted in EPA's doing more economic analysis than other health and safety agencies. In 1996, EPA had 116 economists on its staff. This is a large number, although it should be noted that the agency's staff also included 1,251 lawyers and 1,096 engineers. (EPA Personnel Office, Nov. 1996, as cited in Morgenstern 1997. Figures are for masters- and PhD-level personnel.)

A significant source of the agency's ambivalence regarding economic data is the absence of any agreed-upon mission for the agency. Neither the pollution control statutes nor any other official document spells out EPA's role in society and government. Whether its mission is to balance environmental and economic considerations or simply to advocate environmental positions and let others worry about the costs is unclear. The environmental statutes take a broad diversity of positions along this spectrum.

FRAGMENTATION WITHIN EPA

EPA was created in 1970 from individual programs and functions housed in six different agencies. (USDI, HEW, CEQ, AEC, Fed. Radiation Council, USDA. See Reorganization Plan No. 3 of 1970, July 9, 1970.) All reorganizations cause confusion and trauma for individuals and agencies, but the creation of EPA was unprecedented in terms of the number and size of disparate agencies brought together under a new organizational structure (Davies and Davies 1975, 108). The legacy of EPA's origins has been perpetuated by the fragmented character of environmental law and by the way in which the agency is organized. History, law, and organization have all combined to prevent a unified approach to environmental problems.

Organizationally, EPA has two kinds of offices: program offices (air, water, solid waste, and toxics) and functional offices (enforcement, policy, administration, and research and development) (see Figure 1). The program offices have always been more powerful than the functional offices, and program offices are the dominant players in the regulatory process. The EPA administrator probably has more control over EPA's constituent units than do some other federal department heads (see, for example, Downs 1967; Heclo 1977), but this control is still quite limited, and program offices have a good deal of functional autonomy.

EPA also is fragmented in another way—in the split between headquarters and the ten EPA regional offices. Half of the agency's personnel are located in regional offices. The on-the-ground functions of the agency, such as reviewing permits and conducting inspections, are performed by regional offices. Headquarters program offices set very specific annual goals for their field counterparts, but the field people work for two bosses—their Washington program chief (for example, the assistant administrator for air or for water) and the regional administrator. Increasingly in recent years, regional administrators have tried to exert control over regional personnel at the expense of control from the headquarters program offices.

Two of the regions—Region One (Boston) and Region Eight (Denver)—have recently reorganized on a more functional basis, both to overcome the more fragmented media approach and to give the regional administrator more control over regional personnel. Thus, for example, the four main program offices in Region One are now Ecosystem Protection, Environmental Stewardship, Site Remediation and Restoration, and Environmental Measurement and Evaluation. Whether the regions will be able to relate successfully to headquarters under this new arrangement remains to be seen.

EPA's fragmentation makes it hard to reach decisions. For any decision of consequence, several different organizational units will be involved. Even in a government where meetings are a way of life, EPA is noted both for the amount of time its people spend in meetings and for the large number of people who attend any given meeting.

Meetings, however, are only a symptom of deeper problems that fragmentation has created. The agency is handicapped in setting rational priorities and in applying risk-based criteria because resources are allocated and agendas set by compartments based on the subjects of EPA's statutes (such as groundwater, nonpoint pollution of surface water, or estuaries). Members of Congress, agency personnel, and interest groups tend to think *within* these arbitrary compartments rather than *across* them. Setting priorities is thus another problem created by fragmentation.

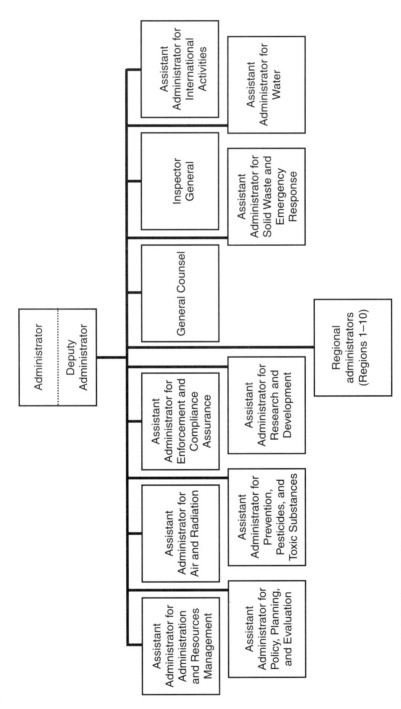

Figure 1. U.S. EPA Organization Chart

Source: U.S. EPA 1996b

ABSENCE OF FEEDBACK AND EVALUATION

EPA has numerous management shortcomings (NAPA 1994), but none is more damaging to the regulatory system as a whole than the absence of feedback and evaluation. This absence means EPA has no reporting system to tell whether its goals are being accomplished, whether any progress is being made, or how much work is being done. Until recently, the agency did have a system that attempted to fill these functions: most recently known as STARS (it went through a variety of acronymic titles), the system was criticized both inside and outside the agency for its heavy emphasis on quantifiable inputs, such as the number of permits issued or the number of enforcement actions initiated. The system was notably lacking in output measures—it said little or nothing about whether environmental conditions were getting better or worse.

Upon becoming EPA administrator, Carol Browner abolished STARS. Initially, no attempt was made to replace it, but efforts to develop a more results-oriented system began in 1996 and 1997. These include establishment of a Center for Environmental Information and Statistics and formulation of a set of environmental goals and milestones.

The agency still has a sub rosa reporting system that has operated for many years between the headquarters program offices and their regional counterparts. Thus, for example, regional people working for the Office of Solid Waste and Emergency Response (OSWER) must report on a monthly basis how they have performed in relation to several hundred targets and measures established by headquarters OSWER staff, such as how many inspections have been conducted and how many permits have been reviewed. These reports are not distributed outside of the particular functional office, and even their existence has been unknown not only to the public but even to many upper-level EPA officials.

The lack of any kind of official reporting mechanism that would tie the agency together or provide EPA's leadership with information about what the constituent parts of the agency are accomplishing is such a glaring omission that it must be remedied before much longer. But the weaknesses of the agency's environmental monitoring system are so great that whatever system is put in place will still have to be based largely on administrative measures, on "bean counting," rather than on actual changes in environmental quality (see Chapter 5).

The lack of any program evaluation capability reinforces the lack of intelligence created by the absence of a regular reporting system. EPA has never encouraged examination of whether its programs are effective, whether they are accomplishing their purpose. For many years, EPA's policy office (which has "evaluation" as part of its title) had a unit labeled the Program Evaluation Division. This division, however, did not perform

critical or independent program evaluation; rather, it served as a management consultant group to EPA program offices.

It does not require a degree in management science to know that it is
important for any organization to have a system for measuring and evaluating what, if anything, it is accomplishing. The absence of such systems
is a major shortcoming of the pollution control regulatory system.

ROLES OF OMB, OSTP, AND CEQ

Three units within the Executive Office of the President—the Office of
Management and Budget (OMB), the Office of Science and Technology
Policy (OSTP), and the Council on Environmental Quality (CEQ)—have
significant decisionmaking responsibilities in the pollution control regulatory process at the federal level.

OMB's primary responsibility is putting together the President's budget submission to Congress. However, preparing the budget is only one
of OMB's functions. Its basic mission is to enforce the policies and views
of the President on the rest of the executive branch. To implement this
mission, it not only makes the budget decisions but also approves all testimony that agencies want to give to Congress, clears all proposed legislation, and referees and frequently decides most interagency disputes. It is
a very powerful organization.

OMB's most visible role with respect to pollution control has been its
review of major regulations. The various executive orders requiring economic analysis of major regulations have usually relied on OMB to
review the analysis. Under some of the orders, the office has been
empowered to delay implementation of the regulation until it is satisfied
with the analysis. OMB's performance of this role has been sharply limited by the small number of OMB reviewers compared with the number
of regulations to be reviewed, by EPA's superior knowledge of the content
and context of the proposed regulation, and by the fact that the legal
authority to issue the regulation resides with the EPA administrator. OMB
has sometimes been effective in getting EPA to change or even abandon a
proposed regulation. (Examples include the proposed Clean Air Act
incinerator recycling requirement withdrawn in 1991 and the 1993
Resource Conservation and Recovery Act landfill rule. See *Inside EPA*
1990a and *Inside EPA* 1993.) Such occurrences are rare, however, and the
OMB review process has not been adequate to ensure that EPA's regulatory impact analyses are of satisfactory quality (see Morgenstern 1997).

The Office of Science and Technology Policy exists to provide scientific advice and coordination to the executive branch. It has been active in
various environmental areas, notably climate change and risk assessment.

Climate change research programs and budgets are coordinated under the U.S. Global Change Research Program. The program is run by the Subcommittee on Global Change Research of the Committee on Environment and Natural Resources of the National Science and Technology Council. The council is a cabinet-level interagency committee chaired by the President and staffed by OSTP.

OSTP also weighs in on a few environmental scientific issues. For example, it took strong exception to a draft EPA report on electromagnetic fields that the agency produced in 1990. D. Allan Bromley, then the director of OSTP, tried unsuccessfully to block publication of the study (*Inside EPA* 1990b; *Inside EPA* 1990c). In the Clinton administration, OSTP has taken an interest in risk assessment, creating a risk assessment subcommittee of the National Science and Technology Council.

The Council on Environmental Quality, created by the National Environmental Policy Act of 1970, has been through a series of ups and downs. Several different presidents have tried unsuccessfully to abolish it. CEQ's symbolic position as the environmental presence in the Executive Office of the President and as the policeman for legally mandated environmental impact statements has kept it alive.

CEQ in theory performs two major types of functions—it serves as a source of environmental policy ideas and it coordinates federal agencies on environmental policy matters. It has, at times, performed the first function quite well. It was the source of many of the landmark environmental laws and policies enacted in the early 1970s. However, CEQ has never been able to perform a coordinating function, because (unlike, for example, OMB) it lacks any leverage over federal agencies. It is not perceived as speaking for the President, it does not control vital functions such as the budget process, and its information resources are vastly inferior to those of the agencies because of its small staff.

OMB, OSTP, and CEQ each get involved in making environmental policy decisions. OMB's involvement with regard to budget matters and regulatory oversight is continuous, as is CEQ's involvement with the environmental impact statement process. For the most part, however, the impact of the three executive office entities is sporadic and limited to occasional high-profile controversies. The output of the pollution control regulatory system is driven by the day-by-day activities of EPA, its state counterparts, and the members of the regulated community.

REFERENCES

Davies, J. Clarence, ed. 1995. *Comparing Environmental Risks.* Washington, D.C.: Resources for the Future.

Davies, J. Clarence, and Barbara S. Davies. 1975. *The Politics of Pollution*, Second edition. Indianapolis: Pegasus.

Downs, Anthony. 1967. *Inside Bureaucracy.* Boston: Little Brown.

Goldman, Lynn R. 1994. Testimony before the Subcommittee on Toxic Substances, Research and Development, Committee on Environment and Public Works, U.S. Senate, May 17.

Heclo, Hugh. 1977. *A Government of Strangers*. Washington, D.C.: Brookings Institution.

HWCP (Hazardous Waste Cleanup Project). 1993. *Exaggerating Risk.* Washington, D.C.: HWCP.

Inside EPA. 1990a. Reilly Fights Major White House Forces for Recycling in Incineration Reg. December 21.

———. 1990b. Bush Science Advisor Blocks EPA Study Linking Electricity to Cancer. December 7.

———. 1990c. EPA Report Linking Electricity to Cancer Released Over White House Objections. December 21.

———. 1993. OMB Rejects Key EPA Landfill Rule, Threatening Cities' Compliance Deadline. May 7.

McCray, Lawrence E. 1983. An Anatomy of Risk Assessment. In National Research Council, *Risk Assessment in the Federal Government, Managing the Press: Working Papers*. Washington, D.C.: National Academy Press.

Morgenstern, Richard. 1997. *Economic Analyses at EPA*. Washington, D.C.: Resources for the Future.

NAPA (National Academy of Public Administration). 1994. *Getting the Job Done.* Washington, D.C.: NAPA.

———. 1995. *Setting Priorities, Getting Results: A New Direction for EPA*. Washington, D.C.: NAPA.

U.S. EPA (Environmental Protection Agency). 1979. Advanced Notice of Proposed Rulemaking. *Federal Register* 44: 60056, October 17, 1979.

———. 1989. Asbestos Ban and Phaseout, Final Rules. *Federal Register* 54: 29460, July 12.

———. 1996a. *Office of Pesticide Programs Annual Report for 1996*, EPA 735-R-96-001. Washington, D.C.: U.S. EPA.

———. 1996b. *EPA Headquarters Directory*. September.

U.S. OMB (Office of Management and Budget). 1997. *Unified Agenda.* http://www.access.gpo.gov./su_docs/aces/aaces002.html.

4

The Federal-State Division of Labor

The two institutional hallmarks of the American governmental system—separation of powers and federalism—are both critical features of the pollution control regulatory system. We have described above some of the effects of separation of powers (see Chapters 2 and 3). Federalism also is a major aspect of the regulatory system, and the states and the federal government share most of the responsibilities for controlling pollution (Lowry 1992; Ringquist 1993).

EVOLUTION OF FEDERAL-STATE RESPONSIBILITIES

State and local regulation of pollution predates federal regulation by many decades. Air pollution, primarily an urban phenomenon, was regulated by city governments, whereas states were the primary regulators of water pollution. Many states were dealing with water pollution problems in the 1920s and 1930s (Michigan Public Act of 1929, No. 245, Michigan Compiled Laws Annotated, Ch. 323; Pennsylvania Laws of 1937, P.L. 1987, Penn. Statutes, Title 35, Ch. 5; General Laws, Rhode Island, 1956, Title 46, Ch. 12-1914 [1920]; West Virginia Laws of 1937, Ch. 130; South Dakota Codified Laws, Title 34A, Laws of 1935, Ch. 174). By 1948, every state had some agency responsible for pollution control (Davies 1970, 38). The modern era of air pollution control began with the efforts of Los Angeles to control smog in the late 1940s and with a 1948 air pollution episode in Donora, Pennsylvania that was held responsible for 20 deaths and almost 6,000 cases of illness (Davies 1970, 34, 49).

Federal involvement in pollution control also began much earlier than is generally realized. The first stream pollution investigations by the federal government began in 1910, and in 1912 the U.S. Public Health Service was authorized by statute to conduct investigations into the pollution of navigable waters (Davies 1970, 25). The Public Health Service investigation of the Donora episode was instrumental in establishing the link between air pollution levels and the deaths and illnesses that occurred there (Davies 1970, 34). During the 1950s and 1960s, the U.S. Public Health Service provided extensive technical assistance to states and localities dealing with air and water pollution.

Evolution toward federal regulatory dominance began in 1956 with enactment of federal authority to curb pollution of interstate waters (Water Pollution Control Act Amendments of 1956, P.L. 84-660). Water pollution enforcement authority was strengthened in 1961 and 1965 (see Davies 1970, 39). Similar enforcement authority was enacted for air pollution in 1963, followed in 1965 by direct federal authority to set and enforce standards for air emissions from new automobiles (Davies 1970, 50). A turning point was reached with the 1970 Clean Air Act and the 1972 Clean Water Act. These acts gave the federal government the dominant role in setting pollution control standards and greatly strengthened the federal enforcement role. Between 1970 and 1994 a large number of legislative enactments steadily ratcheted up the federal regulatory authority (Chapter 2).

The increasing legal power of EPA was balanced by the agency's increasing delegation of program implementation to the states. The air, water, and solid waste acts all allow EPA to delegate permitting and enforcement to states, provided that a state meets certain requirements assuring its implementation capability. As of 1994, for example, forty-six states had been delegated responsibility for implementing basic functions under the Resource Conservation and Recovery Act. Thirty-eight states had authority to issue water pollution discharge permits. Thirty-nine states had full authority and the other eleven partial authority to implement new source performance standards under the Clean Air Act (NAPA 1995, 192-7).

Another turning point was reached in 1995 with the Republican electoral sweep. Republicans, philosophically and politically more pro-business and anti-regulation than Democrats, had run on a commitment to curb "excessive" federal regulation, and many regarded environmental regulation to be the most egregious example of such excess. The House majority whip likened EPA to the Gestapo (Tom Delay quoted in Morgan 1995, A1).

The alienation of the states by the mid-1990s was as important as the Republican capture of Congress. This evolutionary change began in the 1980s. The states generally had supported the federal pollution control role throughout the 1970s. Although friction always existed, states appreciated having a federal enforcement presence—what EPA Administrator Ruckelshaus referred to as "the gorilla in the closet" that states could use as a threat against polluters. States began to resent federal oversight of their actions as they became more competent and professional, as federal funding became less important to state pollution control agencies, as political currents increasingly favored decentralization, and as the most egregious pollution problems were brought under control and attention turned increasingly to small dispersed sources. State and local resentment was also fueled by federal imposition of "unfunded mandates," often-expensive requirements that sometimes were perceived as serving no useful purpose and for which the federal government did not provide adequate funding (see below).

EPA's regional offices fed the state appetite for independence. Despite numerous EPA declarations that the states were partners and should be treated as such, regional offices often treated state officials as incompetent subordinates. A 1995 GAO study found that 63 percent of state environmental managers surveyed regarded the level of EPA control over the states as a significant barrier to program implementation (U.S. GAO 1995, 42). Eighty percent of the managers said that EPA needed to do a better job of consulting the states on key issues (U.S. GAO 1995, 47–48). By 1995, states were demanding an end to EPA's close oversight.

A new approach was outlined in the May 1995 announcement of a "National Environmental Performance Partnership System" agreed to by an EPA-state task force (*Inside EPA* 1995). The proposed system would have seven principal components: increased use of environmental goals and indicators, new approaches to program assessments by states, environmental performance agreements, differential oversight, performance leadership programs, public outreach and involvement, and joint system evaluation.

The core concept of the system was reduced EPA oversight based on states' actual performance in improving the environment. Unfortunately, the performance indicators agreed to are not really adequate to judge performance (Florida State University 1995). As of April 1997, twenty-four states had performance partnership agreements with EPA (Kent 1997).

Parallel to initiation of the performance partnerships, EPA established "performance partnership grants" to allow states more flexibility in their use of grant funds received from the agency. EPA made funds available for multimedia or other purposes not strictly within the traditional grant programs. Several states (such as New Jersey and Delaware) have used the more flexible funds for pollution prevention and other innovative programs. However, most of the performance partnership grants have been used for administrative flexibility (such as combining reporting requirements for several programs) rather than for actually doing different work. Although the grants do not require the receiving state to have a performance partnership agreement, states are required to do the same analysis for administrative as for programmatic shifts in the grant funds.

EPA is hopeful that over time more states will use the flexibility for programmatic innovations (Kent 1997).

CURRENT STATE CAPABILITIES

All observers agree that the capability of state environmental agencies has improved significantly since 1970. Improvement in the professionalism and competence of state governments generally has been reinforced by the increased importance of environment as an issue. Total state expenditures on air quality increased from $80 million in 1971 to $226 million in 1986 to $542 million in 1994 (1971 figures are from Davies and Davies 1975, 161; 1986 and 1994 figures are from CSG 1996, 126). State expenditures on water quality went from $52 million in 1972 to $976 million in 1986 to $1.9 billion in 1994 (Davies and Davies 1975, 155; CSG 1996, 126; figures may not be completely comparable). The increase is impressive even when inflation is taken into account. In constant 1992 dollars, air quality expenditures went from $249 million in 1971 to $516 million in 1994; water quality expenditures went from $155 million to $1,809 million.

Lowry (1992, 5–6) cites a number of reasons for the revitalization of state government: representation in state legislatures has become more equitable since the Supreme Court's decision in *Baker v. Carr* (369 U.S. 186, 1962); public participation in state policy processes has increased; a number of states began to run fiscal surpluses in the 1980s as a result of an improved national economy and increased state taxes; state officials' records are more visible and subject to media attention, especially if they are considering running for a higher office; and the development of state political institutions has fostered policy competition among the states.

Despite a significant general improvement, a very wide range of resources and ability is evident among the states. Total state environmental expenditures in 1994 ranged from $2.1 billion in California to $28 million in Hawaii, almost a hundredfold difference (CSG 1996, 123). Per capita expenditures ranged from $338 in Alaska to $18 in Ohio; California spent $66 per capita, whereas Michigan spent only $20 (CSG 1996, 124). The number of employees in state environmental agencies ranged from 4,486 in California and 4,245 in Florida to 136 in North Dakota and 167 in Nevada (CSG 1996, 4–103).

The wide range of state pollution control resources reflects in part a wide range of cultural and political factors that influence a state's effectiveness in controlling pollution. Political factors, such as the extent of business influence within the state, frequently affect both the stringency of requirements in air and water effluent permits issued to a facility and the amount of effort put into enforcing the requirements. Such factors may even affect whether a permit is issued at all. In Texas, older facilities "grandfathered" into the pollution control system are exempted from needing a permit. Such nonpermitted facilities produced more than half the industrial air pollution in Texas in 1994 (Dawson 1997, 1). The GAO found wide variations in the conditions of permits that states issued to municipal wastewater treatment facilities (U.S. GAO 1996).

There are other prominent indicators of the failure of states to take necessary action. Nonpoint sources of water pollution, such as runoff from farms and roads, are the major current source of water pollution. The federal government has left regulation of these sources largely in the hands of states, but almost without exception the states have not taken adequate control measures. The politics of regulating nonpoint sources is difficult (it is politically easier to levy a fine against a large corporation than against a small farmer), which partly explains federal inaction. Inaction by the states, however, is a potent reminder that decentralization is no panacea.

Some variation in standards and control measures is desirable. Barring any offsetting negative effects on other jurisdictions, a state or locality should be allowed to make its own decisions about how clean it wants the air or how much money it wants to invest in water purification. But there are *always* some effects on other jurisdictions; also, individuals within a community may be entitled to some minimal protection (such as. breathable air). There may be local conditions that do not raise these interjurisdictional problems. North Dakota sets air pollution standards for sunflower seed processing plants, a major potential pollution source in that state. Most other states do not need to deal with this problem.

As states have assumed more responsibility, and as relations between the states and the federal EPA have worsened, EPA has increased the visibility of its state oversight. The agency intervened in a major enforcement case in Virginia, EPA Administrator Browner declaring that the state showed "a lack of leadership" and a "disregard for the requirements of Federal laws" (Cushman 1997, A22). Similar clashes have occurred between EPA and Rhode Island and Pennsylvania.

The conflict between EPA and the states has broadened into general charges and countercharges. Browner has been quoted as saying that "a number of states … are retreating from their commitment to enforce the laws" (*New York Times* 1996, 1), prompting a reply from the association of state environmental commissioners that they were "very concerned about what appears to be a retreat on your part from the partnership relation" (*State Environmental Monitor* 1997, 4).

Political scientist DeWitt John (1993) and others have argued that the real initiative in environmental policy now lies at state and local levels.

John's concept of "civic environmentalism" describes a system of state and local initiatives that is focused on problems involving numerous dispersed sources (such as household garbage and agricultural use of fertilizers) using nonregulatory tools and encouraging a cooperative rather than confrontational approach. He argues that these problems and approaches will dominate the coming years.

Historically and at present, many of the most exciting innovations in environmental policy are at state and local levels. New Jersey, for example, has a bold pollution prevention program that includes facility pollution prevention plans, integrated permitting, and materials-balance monitoring (see Beardsley 1996; Beardsley and Davies 1997). Southern California is implementing a major initiative utilizing market trading schemes to control air pollution (see NAPA 1994). Several states are experimenting with unified permitting and inspections and with innovative reporting requirements. Among the large number of experiments in new forms of citizen participation are the Clark Fork Basin Steering Committee in Montana (see Snow 1996), which has used consensus methods to settle some very divisive water issues, and the North Dakota Consensus Council, a private corporation supported by public and private leaders that encourages discussion and resolution of issues (see Levi and Spears 1994).

The incentives for states to innovate and their ability to do so varies among pollution control programs. Lowry (1992) has shown that the federal government's degree of involvement and the potential for interstate competition are crucial dimensions in determining how much state leadership will be in a program. The less the extent of federal involvement and the lower the possibility of interstate competition to attract pollution sources, the greater the potential for state leadership. However, the match between strong state programs and the severity of a problem will be best when federal intervention is high and interstate competition is low (Lowry 1992, 89). Also, high federal involvement leads to increased dissemination and coordination of state efforts (Lowry 1992, 126–27).

CRITERIA FOR DIVISION OF LABOR

What should be the division of labor between the states and the federal government for administering pollution control regulatory responsibility? There are several ways to address this question. (For an insightful discussion, see Patton 1996.)

Perhaps the least useful way to address the state-federal allocation is through constitutional analysis. The entire Tenth Amendment to the Constitution says: "The powers not delegated to the United States by the Constitution, nor prohibited by it to the states, are reserved to the states respectively, or to the people." For more than fifty years, the prevailing opinion in the courts has been that the power of the federal government is not constitutionally constrained by any powers reserved to the states. In a 1941 opinion (*U.S. v. Darby*, 312 US 100, 61 S. Ct. 451, 85 L. Ed. 609), the Supreme Court stated: "Our conclusion is unaffected by the Tenth Amendment. The amendment states but a truism that all is retained that has not been surrendered. There is nothing in the history of its adoption to suggest that it was more than declaratory of the relationship between the national and state governments as it had been established by the Constitution before the amendment...." In other words, the Tenth Amendment is not a limitation on what the federal government can do.

In recent years, however, a more conservative Supreme Court has begun to hand down a line of decisions placing curbs on federal power. In a 1992 decision, *New York v. United States*, the Court said Congress had violated the Tenth Amendment by requiring New York and other states to take possession of low-level radioactive waste generated within their borders by hospitals and power plants. "The federal government may not compel the states to enact or administer a federal regulatory program," declared the decision. In 1995, the Court for the first time in sixty years invalidated a federal law on the grounds that Congress had exceeded its authority under the Constitution's commerce clause (*U.S. v. Lopez*, 63 LW 4343 [1995]). In 1996, the Court further bolstered states' rights when it ruled that the Eleventh Amendment of the Constitution prevented Congress from giving Indian tribes the right to sue states in federal court (*Seminole Tribe of Florida v. Florida* 3/27/96; for general discussion see Moore 1995). The Court continued the line of reasoning begun in 1992 with a five-to-four decision invalidating a portion of the Brady gun control act on the grounds that the Constitution prohibited the federal government from requiring states to implement a federal regulatory program (*Printz v. U.S.* and *Mack v. U.S.*).

The Supreme Court's shifts will undoubtedly influence what Congress and the states do. However, as changes over the past sixty years indicate, the Constitution is a flexible document—its interpretation will be influenced by the general political climate and the views of individual justices. A discussion of what the federal-state division of labor *should* be must start elsewhere.

A nonlegal analysis based on which pollution problems cross state lines and which do not is somewhat more helpful—only "somewhat" because almost all pollution problems have some kind of interstate dimension. Drinking-water systems serve interstate travelers; waste dumps may contaminate interstate aquifers. The federal government must play a role in dealing with global problems, such as stratospheric ozone depletion, and with problems that have significant interstate dimensions. Voluntary deference or good will is not likely to provide adequate protection for states downwind or downstream from major pollution sources in other states.

Economics and politics provide other criteria. It is generally more efficient for the federal government to regulate products with national markets. Often the political pressure for federal action and national standards comes not from environmental groups but from the manufacturer of the product to be regulated. The manufacturer does not want to have to meet fifty different standards and deal with fifty different government agencies. But politics cuts in various ways. From an environmental standpoint, it may not be realistic to expect a state to regulate one of its major sources of employment, or for a state to have the power and resources necessary to control the actions of a multinational corporation.

Political responsiveness is often cited as a criterion for allocating responsibilities. Those favoring decentralization argue that the government "closest to the people" or "closest to home" should be given authority. Responsiveness is an important consideration, but it is not clear that the smaller and geographically closer governments are always the ones most responsive. Many states and local governments are less open to public participation than the federal government. In many states it is extraordinarily difficult even to find out how a legislator voted on an issue. State and local governments are more subject to the influence of economically important local constituents (such as a major local employer), are generally less open in sharing information than the federal government, and are less familiar to many constituents in these days of national media prominence.

Ultimately, the most fundamental criterion may simply be capability: which level of government can most effectively and efficiently perform a particular function. Using this criterion, most informed observers conclude that for most pollution control programs the federal government should set minimum national standards, with some flexibility for local variation. Permitting, inspection, enforcement, monitoring, and other on-the-ground implementation functions should be done by the states, with some backup and residual legal authority resting with the federal government. (NAPA 1995 does a good job of describing this allocation.)

Any workable federal-state allocation of pollution control functions will be complex and will involve a good deal of interaction and information exchange among the different levels of government. Any simple solution that involves complete and absolute assignment of functions to one level of government is likely to be inefficient, ineffective, or both.

UNFUNDED MANDATES

A major irritant to states and localities has been the imposition of very expensive federal requirements that also may seem to be unproductive uses of local resources. The different aspects of this problem need to be considered separately.

If national requirements are not a good use of taxpayers' money, then regardless of who is paying the money should not be spent. EPA requires the state of Wisconsin to monitor its drinking water for radionuclides even though these contaminants have never been present in that state's drinking water and there is no reason to believe that they ever will be. EPA does not dispute that this requirement is a waste of money; rather, it argues that its regulations do not allow waivers from radionuclide monitoring (Guerrero 1995, 12–13). The solution is obvious: EPA should amend its regulations and give Wisconsin a waiver.

A more difficult problem occurs when the proposed expenditure is useful but is a low priority for the local government and high priority for the federal government. This is a much more typical situation than the Wisconsin radionuclide problem, and it raises the larger question of who should pay for pollution control. Examples include situations where the local government is producing pollution but not feeling its effects (exporting it to some other jurisdiction). Controlling acid rain, a high priority for the federal government because of its effects on the northeastern states and on Canada, was low priority for the midwestern states where much of the acidic emissions originated. Other disparities in priorities occur because the federal government does not carefully consider the priority of environmental expenditures over other kinds of expenditures. Such tradeoffs are much more apparent at the local level: a community that has severe problems with crime, education, and keeping the streets paved may not rate air pollution problems as high priority.

The accepted general principle is that the polluter should pay for pollution control (see Rehbinder and Stewart 1988, 227–28). However, this principle is less helpful than it might seem. In some situations, such as abandoned waste sites, the polluters cannot pay because they have long since disappeared. In other situations, it is politically or economically undesirable to make the polluter pay. Then there is the question of whether costs assessed to the polluter are in fact borne by the polluter. Except in highly competitive markets, polluters often can pass the costs of pollution on to customers without any loss of profit. Even in competitive markets, if all suppliers produce approximately the same type and amount of pollution, the cost will be passed on in the form of price increases.

The polluter-pays principle implies that state and local governments should deal with the unfunded mandates problem by assessing taxes or charges to the polluters sufficient to cover the mandated costs. Often this policy is at least a partial solution, but it runs up against a variety of real-world constraints. State and local governments are often limited in their taxing power by legal constraints enacted by taxpayers. Even in the absence of legal constraints, the governments may be reluctant to impose charges or taxes because of a fear that it will impair their ability to attract new business and industry.

For the foreseeable future, neither EPA nor the states will have enough money to implement all of the legally required pollution control functions. The public will not be willing to pay more taxes and legislators will not be willing to allocate enough existing tax money to environmental functions. The General Accounting Office found that "significant backlogs of expired NPDES [water quality] permits have accumulated in some states, while many states are also having difficulty monitoring environmental quality, setting standards, and enforcing compliance" (U.S. GAO 1995, 3). The shortage of funding reinforces the need to set priorities more effectively.

Having described some of the institutions and processes that comprise the pollution control regulatory system, we next turn to an evaluation of the system using a variety of criteria. The first criterion is whether the system is meeting its own explicit goals—is it reducing the amount of pollution in the environment?

REFERENCES

Beardsley, Daniel P. 1996. *Incentives for Environmental Improvement: Assessment of Selected Innovative Programs in the States and Europe.* Washington, D.C.: Global Environmental Management Initiative.

Beardsley, Daniel P., and Davies, J. Clarence. 1997. Improving Environmental Management: What Works? What Doesn't? *Environment.* 39(7): 6–9.

CSG (Council of State Governments). 1996. *Resource Guide to State Environmental Management.* Fourth edition. Lexington, Kentucky: Council of State Governments.

Cushman, John H. Jr. 1997. Virginia Seen as Undercutting U.S. Environmental Rules. *New York Times,* January 19, p. A22.

Davies, J. Clarence. 1970. *The Politics of Pollution.* Indianapolis: Bobbs-Merrill.

Davies, J. Clarence, and Barbara S. Davies. 1975. *The Politics of Pollution.* Second edition. Indianapolis: Pegasus.

Dawson, Bill. 1997. Pollution Tied to Lack of Permits. *Houston Chronicle.* April 1, p. 1.

Florida State University. 1995. *Prospective Indicators for State Use in Performance Partnership Agreements.* Tallahassee: Florida Center for Public Management.

Guerrero, Peter F. 1995. *Environmental Protection: Current Environmental Challenges Require New Approaches.* May 17. GAO/T-RCED-95-190. Washington, D.C.: U.S. GPO.

Inside EPA. 1995. State-EPA Accord Dramatically Scales Back Federal Oversight. Special report, June 5.

John, DeWitt. 1993. *Civic Environmentalism.* Washington, D.C.: Congressional Quarterly Press.

Kent, Charles. 1997. Telephone interview with the authors, April 21. Kent is in the Office of the Administrator, U.S. EPA.

Levi, Bruce T., and Larry Spears. 1994. Public Policy Consensus Building: Connecting to Change for Capturing the Future. *North Dakota Law Review.* 70(2): 311–51.

Lowry, William R. 1992. *The Dimensions of Federalism: State Governments and Pollution Control Policies.* Durham, North Carolina: Duke University Press.

Moore, W. John. 1995. Pleading the 10th. *National Journal.* July 29: 1,940–44.

Morgan, Dan. 1995. Republicans Defect to Kill Curbs on EPA. *Washington Post,* July 29, pp. A1, A9.

NAPA (National Academy of Public Administration). 1994. *The Environment Goes to Market: The Implementation of Economic Incentives for Pollution Control.* Washington, D.C.: NAPA.

———. 1995. *Setting Priorities, Getting Results: A New Direction for EPA.* Washington, D.C.: NAPA.

New York Times. 1996. States Neglecting Pollution Rules, White House Says. December 15, p. 1.

Patton, Vickie L. 1996. A Balanced Partnership. *The Environmental Forum.* May/June: 16–22.

Rehbinder, Eckard, and Richard Stewart. 1988. *Environmental Protection Policy.* New York: Walter de Gruyter.

Ringquist, Evan J. 1993. *Environmental Protection at the State Level: Politics and Progress in Controlling Pollution.* Armonk, New York: M.E. Sharpe.

Snow, Donald. 1996. River Story. *Chronicle of Community* 1(1): 16–25.

State Environmental Monitor. 1997. State Commissioners Challenge Browner View of State Enforcement. January 6, p. 4. Washington, D.C.: Inside Washington Publishers.

U.S. GAO (General Accounting Office). 1995. *EPA and the States: Environmental Challenges Require a Better Working Relationship,* GAO/RCED-95-64. Washington, D.C.: U.S. GAO.

———. 1996. *Water Pollution: Differences Among the States in Issuing Permits Limiting the Discharge of Pollutants* GAO/RCED-96-42. Washington, D.C.: U.S. GAO.

PART II:
Evaluating the Regulatory System

5

Reducing Pollution Levels

A logical starting point to evaluate the efficacy of the regulatory system is to ask whether the system has reduced levels of pollution released to the environment. However, the question is not as straightforward as it first appears. In theory, EPA programs should be accompanied by information on pollution trends in order to help environmental managers monitor how well they are working. Ideally, environmental managers should possess data that not only show how much pollution is emitted and concentrated in the environment, but also information that illustrates the potential adverse impacts of pollution on people and other living things (see Figure 1). To date, no such comprehensive information system has been developed.

The data deficit is partly a function of the fragmentary nature of the individual air, water, waste, and toxics laws developed to control pollution and partly a result of the complicated nature of pollution. Primarily, monitoring pollution is difficult because substances often travel through more than one medium and may interact with other forms of contaminants to adversely affect human and environmental health in ways that are difficult to disentangle (Russell 1990, 243–74). For example, it was once thought that most nitrogen in water bodies such as the Chesapeake Bay flowed from fertilized farm fields. Better monitoring efforts have enabled scientists to determine that up to 35 percent of nitrogen deposits to the bay originate from air pollution sources such as automobiles (Dennis 1995, 2–3).

Policymakers have long been cognizant of the complexity of controlling, as well as simply characterizing, pollution (Davies 1970, 17–21). As used here, the term "pollution" applies to everything from byproducts that arise from metals and organic chemicals that are used in industrial production to biological oxygen demand in sewage. Unfortunately, media-specific programs are often designed more to ease implementation of fragmented laws than to closely match the actual physical characteristics of the pollution they target. As a result, federal environmental monitoring data typically reflect a bias toward administrative simplicity. Most

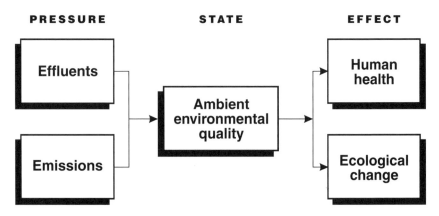

Figure 1. A Framework to Assess the Regulatory System.
Source: U.S. EPA 1995a, 6.

existing efforts to measure progress are comprised of program measures, such as compliance and enforcement rates or permits issued, instead of data based on actual environmental quality (NAPA 1995, 164).

Furthermore, when data based on environmental monitoring are available to evaluate environmental performance, it often is impossible to use the information to track long-term, national trends. Different collection and measurement methods employed by state and local agencies make it difficult, if not impossible, to aggregate data to compare trends over time, or to develop a national profile on the state of the environment (NSTC 1997, 70, 73).

Finally, while it is possible to track pollution trends in some media, it is hard to know the extent to which laws are responsible for the patterns. This conundrum, known as the "causality problem," makes evaluation of pollution control among the most challenging assessment exercises (see, for example, Wholey and others 1970). In the case of pollution, determining causality is complicated by the fact that pollution also is a function of many other factors, including economic activity, weather patterns, and population. For example, MacAvoy found that changes in the emissions patterns of particulates and carbon monoxide were more a function of business investment and coal use than regulations (MacAvoy 1987). A study we conducted found that regulation did have an impact (see below). Overall, it is impossible to document the extent to which regulations have improved environmental quality.

A number of initiatives to improve the quality of environmental data are under way. The most prominent efforts are aimed at improving environmental monitoring and research across government (NSTC 1996). Others, such as the Intergovernmental Task Force on Monitoring Water Quality (ITFM 1994), are focused around specific media such as water, where data are exceptionally poor. Those initiatives trained specifically on improving the quality of pollution control data typically seek to improve certain categories of information or collection and reporting methods (U.S. EPA 1995a). For example, EPA's Environmental Monitoring and Assessment Program primarily is designed to improve the quality of ecological indicators that illustrate the effects of pollution on plants and animals.

OVERVIEW

Given the information deficiencies, the data that do exist—most notably for air and for some rivers and streams—show that the amount of pollution released to the environment has declined since the 1970s. It should be noted that the declines coincided with substantial increases in the size of the nation's population, economy, and activities such as vehicle use that contribute to pollution. Between 1970 and 1995, the total U.S. gross domestic product increased 99 percent; population climbed 28 percent; and vehicle miles traveled jumped 116 percent (U.S. EPA 1996c, 3). While we cannot definitively link the decline in pollution to laws, it coincides with the expansion of the federal regulatory system. The most notable decreases are in emissions to air and overall concentrations of pollutants in the air. In the case of substances such as lead, which has been gradually phased out of use through a variety of policy instruments, declines are pronounced. As we illustrate in the part of the chapter that examines toxics trends, the declines in lead emissions are accompanied by a drop in lead production in the United States as well as levels of lead in human blood. Similar declines are noticeable for other toxic substances and pesticides such as DDT and PCBs, as well as for pollutants such as phosphorous, which have been gradually phased out of use in commercial products (including detergents). While the data register declines in some of the most harmful substances, no systematic inventory exists with which to estimate how much of these substances remains in the natural environment. It is known that substances such as PCBs tend to persist in soil deposits.

The long-term record for pollution to land and other water bodies such as coastal areas, lakes, and parts of rivers and streams located away from monitoring stations is harder to interpret because few national trend data exist. However, it appears that pollution concentrations in rivers and streams have decreased over time. Gains are mostly attributed to pollution problems that stem from stationary sources, such as sewage treatment plants. Problems from diffuse sources such as urban and agricultural runoff persist and promise to be more challenging to solve (Smith,

Alexander, and Lanfear 1991, 127). Similarly, the data show that the pollution control system has made less progress on problems such as municipal solid waste and some classes of emissions such as nitrogen oxides that are linked to consumption and transportation patterns in the United States (U.S. EPA 1995c, 2).

This chapter presents in greater detail pollution monitoring data developed primarily by federal agencies charged with pollution control. When possible, we examine data since 1970 that coincide with the expansion of the federal regulatory system in order to assess whether pollution levels have dropped over the period. However, as we discuss at the end of this chapter, declines in pollution levels that have followed increases in regulatory efforts are not necessarily caused by regulatory efforts.

This chapter presents the pollution record for air, water, wastes and toxics in descending order of data availability. For most media, data exist that estimate emissions trends. For air and water, some monitoring networks provide a limited picture of national air and water quality. Data that link pollution to environmental and human health effects are in shorter supply. To date, information on long-term emissions trends, air quality, and human and environmental health effects of air pollution are the most comprehensive. Data that show long-term, national hazardous waste and toxic trends must still be developed.

While most of the data are from other sources, the chapter concludes with a summary of our own efforts to statistically isolate the effects of air pollution control laws on pollution levels in three major metropolitan areas in the United States.

AIR POLLUTION

In general, data on emissions and air quality show that technical controls have helped to stem air emissions from cars and factories (U.S. EPA 1995c, ES-2). While data on air emissions provide estimates on what materials might be coming out of a smokestack, decisionmakers historically had no way of knowing where conventional pollutants ultimately end up. Some preliminary estimates on the effects of air pollution show that while pollution loads to the environment have dropped or stabilized, scientists are just beginning to document the potential effects of air pollution on forests, water bodies, and other natural resources.

In order to interpret air quality and emissions trends, it is helpful to understand what types of pollutants federal legislation targets. Congress passed the Clean Air Act of 1970 to provide air quality sufficient to protect human health and welfare. The statute directs EPA to set primary standards that "provide an ample margin of safety" to protect public health and to develop secondary standards to protect welfare (42 U.S.C. 7400). The statute requires EPA to establish acceptable levels of concentration or "criteria" in the outside (ambient) air for five pollutants: carbon monoxide (CO), particulate matter, nitrogen oxides (NO_x), volatile organic compounds (VOCs), and sulfur dioxide (SO_2). Lead was added under a later version of the act, and ozone was adopted as the indicator for VOCs. The particulate matter standard changed in 1987 to focus on small particles (PM-10), and again in 1997 to focus on still smaller particles.

These concentration levels are referred to as the "NAAQS" or National Ambient Air Quality Standards. The Clean Air Act mostly targets criteria pollutants from two groups: stationary sources, such as existing and new industry, and mobile sources, such as highway vehicles. The act primarily charges states with achieving air quality goals.

Emissions

EPA has developed estimates of pollution emissions to air from 1900 onward. The only exception is data on fugitive dust, which EPA added to the PM-10 category after the standard changed in 1987 to focus on that category of particles that is smaller than ten micron widths in diameter. EPA has prepared estimates on fugitive dust sources back to 1985. In 1995, fugitive dust accounted for about 92 percent of PM-10 emissions (U.S. EPA 1996c, 15).

Released and revised annually, EPA's emissions estimates are derived from engineering calculations rather than actual monitoring devices. Furthermore, estimates apply primarily to emissions generated by humans and, to a lesser extent from natural sources such as vegetation, livestock, and biological decomposition. Despite their limitations, the EPA emissions data provide the only long-term national profile of air pollution emissions patterns.

EPA estimates that emissions doubled and, in some cases, trebled over the course of the century, peaking sometime between 1965 and 1975. For example, SO_2 emissions increased from about ten million tons in 1900 to a peak of around thirty million tons in the 1970s. VOC emissions similarly increased from around seven million tons to well over twenty-five million tons during the late 1960s. NO_x shows similar long-term increases. However, unlike SO_2 and VOCs, nitrogen oxide emissions have not dropped significantly since 1970 (U.S. EPA 1996c, 6–7).

Figure 2 illustrates emissions trends since 1970. All substances except nitrogen oxides have declined over the period (U.S. EPA 1996c, A2-A27; 1996d, 2). Declines primarily stem from improved methods to control emissions from combustion sources. Cleaner industrial and automotive fuels and improved control technology have helped to cut emissions from

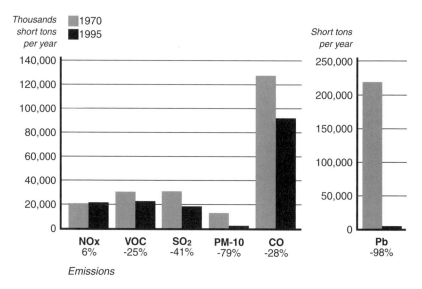

Figure 2. National Air Pollution Emissions Trends, 1970–1995.

Source: U.S. EPA 1996c, A2–A27.

smokestacks and tailpipes. EPA cautions, however, that spiraling automobile use may threaten to eventually outstrip the gains.

Lead emissions, formed primarily from the use of leaded gasoline, show the most dramatic drop (-98 percent). EPA reports that lead emissions dropped precipitously primarily due to regulations that reduced the lead content of gasoline. In contrast, nitrogen oxides, which contribute to the formation of nitrogen dioxide, increased by roughly 6 percent over the period. EPA attributes the increase to emissions from electric utility fuel combustion and vehicles. Without highway vehicle standards, EPA estimates that nitrogen oxide emissions would be more than double current emissions.

When fugitive dust is factored out, particulate matter emissions from industrial activity and fuel combustion have declined 79 percent since 1970. Fugitive dust, which is sensitive to wind and weather patterns, has decreased about 17 percent since 1985, the first year for which EPA prepared estimates of the emissions category. Particulate matter declines are followed by a 41 percent drop in sulfur dioxide emissions since 1970. According to EPA, declines are due in part to improved emissions controls and cleaner fuels adopted by the nation's electric utilities.

Carbon monoxide emissions peaked in 1970 and have declined by about 28 percent since then. Vehicles comprise the largest major source of carbon monoxide emissions. Carbon monoxide emissions fell even though total vehicle miles traveled more than doubled between 1970 and 1995. Without programs to stem vehicle emissions, EPA estimates that carbon monoxide emissions would be three times higher today. Such gains in controlling these emissions have been achieved in part through the retirement of older, less fuel-efficient vehicles. EPA notes that the retirement of older vehicles has reached the limit of its effectiveness in the reduction of carbon monoxide emissions. Continued progress may require more challenging strategies to encourage people to change their driving habits and use transportation that generates fewer emissions.

Volatile organic compounds are typically linked to urban smog. Ground-level ozone is formed when sunlight reacts with volatile organic compounds and nitrogen oxide. Since 1970, VOC emissions have dropped by about 25 percent. As with fugitive dust, meteorological conditions complicate efforts to control volatile organic compounds that contribute to ozone formation. Because sunlight contributes to its formation, ozone tends to peak during the summer in most major metropolitan areas. Despite strict control efforts, the substance tends to form in exceptionally high levels during dry, hot years. While VOC emissions have dropped over the period, spiraling automobile use promises to make VOC control increasingly challenging.

Ambient Air Quality Trends

Data on ambient concentrations are a more useful indicator of environmental quality than emissions estimates because they measure actual pollutant concentrations in the air. While ambient data reveal trends in air quality, they cannot be used to pinpoint with precision what underlying factors—climate, industrial output, regulation—drive the trends. EPA reports ambient air quality data as national averages, based on hourly, quarterly, or annual measurements of criteria pollutants at between roughly 200 and 800 monitoring stations. Stations are located in metropolitan areas on the assumption that cities represent places where the highest pollutant concentrations and the greatest amount of human exposure occur.

Because ambient concentrations are reported as national averages, they fail to reflect regional variations in air quality. Ambient data also fail to reflect how other variables such as climate, geography, and industrial composition affect air quality. For example, as the VOC emissions trends illustrate, ozone averages may be more a function of summer temperatures than regulatory controls. Because the number and location of monitoring stations tend to change over time, EPA reports ambient trend data in rolling ten-year intervals that are updated and revised each year. Data more than ten years old are not directly comparable with the ten-year

Table 1. Ten-Year Air Quality and Emissions Trends, 1986–1995.

	Air quality change	Emissions change
Carbon monoxide	–37%	–16%
Lead	–78%	–32%
Nitrogen dioxide	–14%	–3% (NO_x)
Ozone	–6%	–9% (VOC)
PM-10*	–22%	–17%
Sulfur dioxide	–37%	–18%

Source: U.S. EPA 1996d, 1.

Note: PM-10 changes are based on 1988–1995 data.

trend information from EPA's most recent air quality report published in 1996. As a result, it is not possible to derive a continuous profile of air quality since 1970 (U.S. EPA 1996d). At best, we can compare data from two ten-year intervals.

Percentage changes in ambient concentrations from 1986 to 1995 are presented along with the ten-year emissions trends in Table 1. The air quality record is fairly consistent with the percentage change in emissions over the last decade. Like emissions, ambient concentrations show the biggest decreases occurring in lead, carbon monoxide, and sulfur dioxide concentrations. Concentrations of lead dropped 78 percent. Concentration of carbon monoxide and sulfur dioxide both dropped 37 percent between 1986 and 1995 (U.S. EPA 1996d, 1).

Figure 3 pairs the trend data from 1986 to 1995 with ten-year data from 1976 to 1985 to give an overall impression of ambient air quality shortly after the expansion of clean air legislation (U.S. EPA 1996d, 76). As mentioned, the trend data for the two periods are distinct and not directly comparable. Trends are reported as standard level concentrations in either parts per million (ppm) or micrograms per cubic meter ($\mu g/m^3$) and are averaged over hours, days, quarters, or years. For example, carbon monoxide (CO) is measured in parts per million (ppm) and averaged over eight hours. Lead is reported in micrograms per cubic meter and reported as a maximum quarterly average. The standard for each criteria pollutant appears on the vertical axis. Data for PM-10 are available from 1988 onward, reflecting the 1987 change in standards.

It is clear that air pollution levels during the past decade were lower than those reported from 1976 to 1985. Overall, air quality appears to have improved significantly since 1976. As in the case of emissions, lead and sulfur dioxide show the most improvement over the period. Both nitrogen dioxide and ozone show less improvement relative to the other criteria pollutants. Still, the long-term trend for pollutants in both periods is improved air quality.

2nd max
8-hr (ppm)

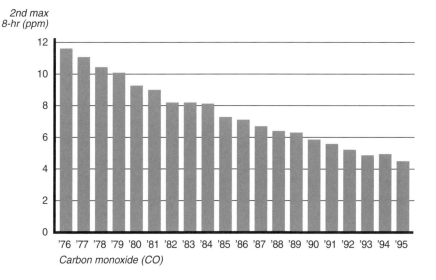

Carbon monoxide (CO)

Arith.
mean
(ppm)

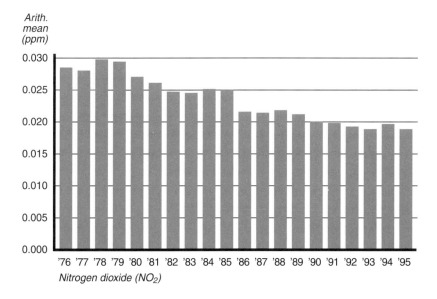

Nitrogen dioxide (NO₂)

Figure 3. National Long-Term Air Quality Trends, 1976–1995: Carbon Monoxide and Nitrogen Dioxide. *(continued on next page)*

Source: U.S. EPA 1996d, 76.

2nd max
1-hr (ppm)

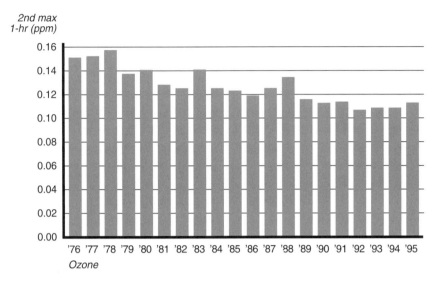

Ozone

Wtd.
arith.
mean
(μg/m³)

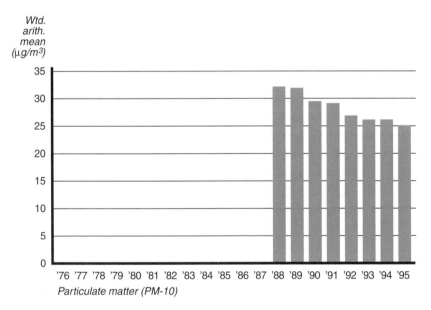

Particulate matter (PM-10)

Figure 3 (continued). National Long-Term Air Quality Trends, 1976–1995: Ozone and Particulate Matter. *(continued on next page)*

Source: U.S. EPA 1996d, 76.

Max. qtr.
(μg/m³)

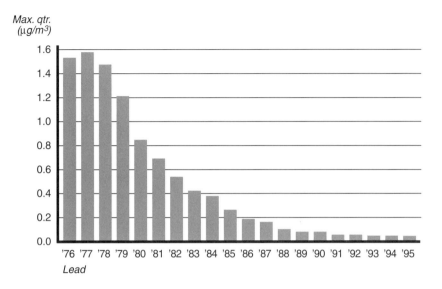

Lead

Arith.
mean
(ppm)

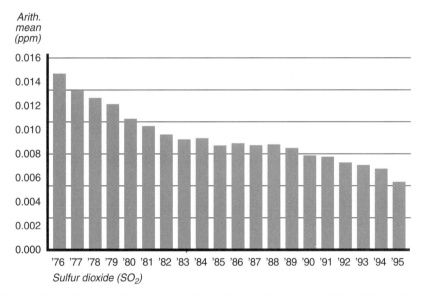

Sulfur dioxide (SO₂)

Figure 3 (continued). National Long-Term Air Quality Trends, 1976–1995: Lead and Sulfur Dioxide.

Source: U.S. EPA 1996d, 76.

Human Exposure

Data on emissions and air quality illustrate pollution trends; but such information alone says little about the potential impact of air pollution on people and on the natural environment. To develop such a profile, EPA pairs air quality with information about where pollution occurs and which populations are potentially exposed. EPA produces two principal measures: a violation measure based on a single year and a formal designation based on data developed over many years. The single-year snapshot shows how many people live in a county where the air quality standard for at least one of the NAAQS was not met. According to the most recent single-year measure, at least eighty million people in the United States lived in such a county during 1995 (U.S. EPA 1996d, 4). While the measure gives one indication of the relative extent of air pollution in 1995, it does not mean that people who lived in a county where one or more of the NAAQS was not met were actually exposed to above-standard levels of that pollutant.

The second measure shows cities that have been subject to formal rulemaking processes under EPA "nonattainment" status. The formal designation is based on multiple years of data to account for changes such as weather that may affect pollution levels. The formal designation data tend to yield larger estimates than the single-year measure of the number of people who live in nonattainment areas. The nonattainment status population figure is typically larger because it often applies to an entire metropolitan area, instead of a single county.

In 1996 (see Table 2), approximately 127 million people lived in counties classified as nonattainment areas (U.S. EPA 1996d, 128). Ozone leads the list of standards not being attained, followed by carbon monoxide. About 1.5 million people lived in areas of nonattainment for lead. It should be noted that EPA formal designations change over time, reflecting air quality improvement, as well as deterioration. As of September 1996, about sixty-eight metropolitan areas were designated by EPA as nonattainment areas for ozone, down from ninety-seven in August 1992. Eighty-one were designated as nonattainment areas for PM-10, up from seventy four years before. Only one city, Los Angeles (see Table 3), is a nonattainment area for nitrogen oxides (U.S. EPA 1996d, 61).

Perhaps the best-known geographic measure developed by EPA reflects air quality trends in major metropolitan areas. The ten-year measure is based on the number of days when air is deemed unhealthy. The measure is based on a pollutant standards index (PSI). PSI index values are reported in all U.S. metropolitan areas with populations of 200,000 or more.

Figure 4 shows the five U.S. cities with the highest number of unhealthful air episodes from 1986 to 1995 (U.S. EPA 1996d, 157). Not sur-

Table 2. Population in Counties Listed as Nonattainment Areas, 1996.

Pollutant	Counties	Population (thousands) (using 1990 census data)
Ozone	68	109,794
Carbon monoxide	31	45,089
PM-10	81	30,943
Sulfur dioxide	43	5,269
Lead	10	1,545
Population in nonattainment areas for one or more criteria pollutant		126,957

Source: U.S. EPA 1996d, 128.

Note: Some areas in the United States are nonattainment areas for more than one pollutant, which is why the total population in nonattainment areas is less than the sum of the population affected by each individual pollutant.

Table 3. Number of Nonattainment Areas for NAAQS Pollutants.

Pollutant	August 1992	September 1996
Ozone	97	68
Particulate matter (PM-10)	70	81
Sulfur dioxide	50	43
Carbon monoxide	38	31
Lead	12	10
Nitrogen dioxide	1	1

Source: U.S. EPA 1996d, 61.

prisingly, the sunny, sprawling cities in southern California top the list. With the exception of Riverside, the number of unhealthful days in Los Angeles is more than double those of the remaining cities on the list. Overall, the number of episodes has declined over time. In Phoenix, the number of unhealthful days dropped from more than eighty-eight to just thirteen during the ten-year period. Severe air episodes in both Ventura and San Diego counties, which border Los Angeles to the north and south respectively, have increased slightly over the past three years.

Human Health Effects

Criteria pollutants can cause problems for healthy children and adults as well as for asthmatics who have impaired respiratory systems. Since the NAAQS were adopted, EPA has drawn from numerous studies that attempt to link exposure to health effects. While such studies can never conclusively show that a pollutant causes illness, scientists use statistics to estimate what concentrations are associated with human health effects.

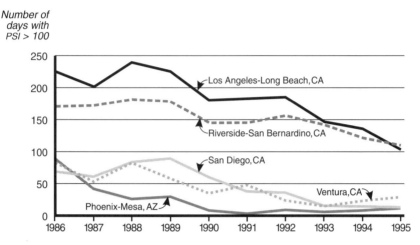

Figure 4. Metropolitan Areas with the Highest Number of Unhealthful Air Quality Days, 1986–1995.

Source: U.S. EPA 1996d, 157.

EPA reports that healthy adults and children exposed to even relatively low concentrations of ozone for six or seven hours can experience chest pain, coughing, nausea, and pulmonary congestion (U.S. EPA 1988). The agency also reports that nitrogen oxide can irritate lungs and lower resistance to respiratory infections such as influenza (U.S. EPA 1995d). Other studies cited by the agency show that high levels of carbon monoxide can impair manual dexterity and learning ability and make it difficult to complete complex tasks (U.S. EPA 1984). PM-10 particles (sometimes in the presence of sulfur dioxide) are suspected by some to impair breathing and respiratory function, aggravate existing respiratory and cardiovascular disease, and alter the body's defense systems (U.S. EPA 1986).

Because scientific methods to assess health effects remain uncertain, and the cost of complying with stringent regulation is high, the actual impact of pollutants on humans remains hotly debated. EPA's recent attempt to revise its criteria for particulate matter is one such example. The agency has sought to identify fine particles (PM-2.5) as a possible indicator for the national ambient air quality standard. Although about fifty epidemiological studies show an association between particulate matter and health effects, many experts believe that EPA's case for a fine particle indictor may be overstated. For the most part, the studies fail to identify the mechanisms or toxic agents in particulate matter that could trigger health problems (U.S. EPA 1996a, 3–4).

Ecological Effects

Emissions and ambient data illustrate trends in total pollutant loads and concentrations in the air but fail to show where pollution is transported and how it may affect the quality of the environment, particularly in parts of the country where certain types of pollution tend to concentrate. To provide a more in-depth assessment of the potential impacts of air pollution, EPA has developed a prototype of broader measures of air quality that attempt to qualitatively link emissions levels to changes that can affect ecological health (U.S. EPA 1994a). At this stage, links between emissions and ecological change are still qualitative because monitoring techniques and ecological risk assessment methods, while improving, are still too crude to make associations with a high degree of certainty.

EPA has developed an inventory that identifies pollutants that commonly end up far away from their source, and shows how they are transported (U.S. EPA 1994a). For example, sulfur dioxide and nitrogen dioxide released from smokestacks and tailpipes react with weather patterns to form the principal components of acid rain. Other pollutants, including volatile organic compounds, tend to migrate from where they originate to regions hundreds of miles away. The data show that parts of the country with the highest concentration of people, such as New York and Los Angeles, tend to have the highest levels of ozone precursors. Dry, sunny regions such as southern California show the highest number of ozone episodes (U.S. EPA 1994a, 11–12). In other parts of the country, pollution patterns are tied to industrial activity. For example, sulfur dioxide emissions are greatest in the East, where use of sulfur-bearing fossil fuels predominates (U.S. EPA 1994a, 10). Sulfur dioxide emissions are much lower in the West, where use of natural gas predominates.

Sulfur dioxide is particularly problematic because it is deposited through rainfall onto land and into water bodies. Consistent with the emissions and ambient data, EPA readings show that sulfur dioxide levels in northeastern lakes and streams declined between 1983 and 1989. On land, the substance can harm some tree species. Researchers have found that acid rain appears to predispose high-elevation red spruce forests in the East to damage from natural stress-causing agents such as pests and disease (U.S. EPA 1994a, 10).

According to EPA's ecological assessment prototype, current ambient levels of ozone generated in metropolitan areas may be harming forests as well. Effects range from damaged leaves to reduced growth, and from increased mortality for plants and trees to wholesale changes in the composition of tree communities (U.S. EPA 1994a, 27). Elevated ground-level ozone in Los Angeles has caused ecosystem changes in pine forests such as in the San Bernardino Mountains, which are located about 100 miles away. Scientists there have observed large-scale effects from pest infestation in certain pine species: they suspect that ozone has helped to make the trees more prone to infestation.

WATER POLLUTION

While extensive data have been developed to assess air pollution trends and the effects of some pollutants on people and other living things, data for water are much less useful (for a good overview, see Knopman and Smith 1993). The data deficit is not due to lack of proper legislation. Created at a time when many of the nation's waterways seemed polluted beyond repair, the Clean Water Act aims to "restore and maintain the chemical, physical and biological integrity" of the nation's water. This ambitious statute applies to more than 3 million miles of rivers and streams, 40 million acres of lakes, 34,000 square miles of estuaries (excluding Alaska), and 58,000 miles of ocean shoreline.

In the act, Congress directed EPA to gauge progress achieved toward cleaning up the nation's rivers, lakes, and coastal areas. Section 305(b) requires states to report to EPA every two years on how well state waters support designated beneficial uses. These uses range from drinking water supply and recreation to support for aquatic life. Attainment grades range from full support of beneficial uses to partial support to nonattainment. EPA collects the Section 305(b) data from states and reports the findings to Congress every two years.

While the 305(b) reports illustrate conditions of some water bodies in some states, they fail to provide a comprehensive picture of national water quality trends. Because the reports cover only a small fraction of water bodies in most states, the areas assessed change over time. The 1994 reports contain data covering only 17 percent of the nation's rivers and streams and 9 percent of all ocean shoreline waters. Such coverage can produce a misleading picture of progress in combating water pollution. For example, of the 9 percent of ocean shoreline waters covered, 89 percent provide full support for beneficial uses. Conversely, of the 98 percent of all Great Lakes shoreline assessed for the 1994 report, only 2 percent provide full support (U.S. EPA 1995d, ES-1–ES-27). (See Table 4.)

In addition to representing just a subset of the nation's water, the data are not comparable among states. Sampling and testing methods employed by state agencies tend to change from year to year and to vary by state. Some data are derived from actual water sampling, while other information is simply derived from "best guesses." Areas assessed also change depending on state finances and other resource constraints.

Table 4. Percent of Surveyed Waters that Support Full Use, 1994.

	Rivers and streams	Lakes, reservoirs, and ponds	Estuaries	Ocean shoreline waters	Great Lakes shoreline
Percent surveyed	17	42	78	9	94
Percent surveyed supporting full use	57	50	57	89	2

Source: U.S. EPA 1995d, ES-1–ES-27, 13.

Table 5. Five Leading Sources of Water Quality Impairment Related to Human Activity.

Rank	Rivers	Lakes	Estuaries
1	Agriculture	Agriculture	Urban runoff/storm sewers
2	Municipal sewage treatment plants	Municipal sewage treatment plants	Municipal sewage treatment plants
3	Hydrologic/habitat modification	Urban runoff/storm sewers	Agriculture
4	Urban runoff/storm sewers	Unspecified nonpoint sources	Industrial point sources
5	Resource extraction	Hydrologic/habitat modification	Petroleum activities

Source: U.S. EPA 1995d, ES-12.

Beginning in 1995, EPA's Water Office began a concerted effort to make the 305(b) reports more uniform and meaningful.

Given these data limitations, the state reports show that while we have made some progress in reducing water pollution, much remains to be done. Of the waters assessed by states for the 1994 report, just 57 percent of rivers and streams and half of the lakes, reservoirs, and ponds "fully support" such activities as fishing, swimming, and water consumption (U.S. EPA 1995d, ES-13, ES-16).

The reports do identify some the most significant factors that affect water quality. For example, the 305(b) reports ask state officials to identify leading sources of pollution (U.S. EPA 1995d, ES-12). (See Table 5.) The state reports mirror data developed through water quality simulation models that show that, while the nation has made great progress in controlling pollution from firms and household sewage, water pollution originating from diffuse sources such as fertilizer from farm fields or oil and grease from city streets remains a persistent problem (Gianessi and Peskin 1981, 1984). Agriculture accounts for the major share of nonpoint discharges, followed by runoff from nonporous surfaces such as roads and parking lots in developed areas. For example, in the Chesapeake Bay, the combined phosphorus and nitrogen loads from nonpoint sources account for up to 44 percent of the total pollutant load (U.S. EPA 1994c, S3–S4).

Effluents

Federal water quality programs focus heavily on efforts to reduce pollutant loads through large investments in public and private control measures. EPA estimates that between 1972 and 1996, taxpayers and the private sector will have spent at least $700 billion on water pollution control, mostly on construction grants for municipal sewage treatment facilities and end-of-pipe industrial controls (U.S. EPA 1990, 8-49–8-50). A logical starting point to assess the Clean Water Act is to examine whether these investments reduced total pollutant loads over time.

To track such trends, it is helpful to understand a few terms. Organic waste controlled by sewage facilities is typically measured in units of biochemical oxygen demand (BOD), which is the amount of oxygen needed to decompose the waste. Left untreated, waste saps oxygen from water and aquatic plants that cleanse water and support marine life. Sewage treatment plants use three tiers of treatment, primary, secondary and advanced, to remove BOD. Primary treatment removes between 25 and 30 percent of BOD, while advanced removes up to 99 percent of BOD.

The Clean Water Act of 1977 greatly expanded requirements that municipal sewage treatment plants attain a level of treatment capacity (secondary) that removes at least 85 percent of several key conventional pollutants, including those that deplete oxygen in water. From 1972 to 1992, superior treatment technology helped reduce by 36 percent the amount of BOD discharged by treatment facilities to water at a time when population and the amount of sewage treated at wastewater facilities rose by 30 percent. In 1992, about 180 million people (out of a total population of 255 million) were served by wastewater treatment systems (U.S. CEQ 1996, 229). While the number of individuals served by advanced treatment levels increased from roughly 4 to 65 million between 1960 and 1988, the number of people not served by wastewater facilities has remained fairly constant over the period, about 70 million (U.S. CEQ 1990, 309).

Water Quality

Data compiled by the U.S. Geological Survey (USGS) on stream-water quality support water quality information from EPA showing declines in pollution from point sources such as industry and sewage treatment facilities. The National Stream Quality Accounting Network (NASQAN) provides the only national profile of pollutant concentration trends in some rivers and streams during the 1970s and 1980s. Compiled by the U.S. Geological Survey, NASQAN data measure concentrations of major dissolved inorganic constituents, nutrients, suspended sediments, fecal bacteria, and trace metals (Smith, Alexander, and Lanfear 1991; Smith, Alexander, and Wollman 1987). NASQAN data are superior to any other source because they are collected by one agency using one sampling method.

While the data are the best available , they still have limitations due to the fact that the network was originally designed not to monitor pollution levels but to measure stream flow. As a result, the stations do not represent a random sampling of surface water quality in the United States. Furthermore, the data do not depict true trends because the number of monitoring stations has changed over time. Most of the 400 network stations are on large rivers, far from point sources such as factories and sewage treatment plants. The stations are therefore unlikely to gauge the level of impact from sources because water has a tendency to "cleanse itself" of some pollutants as it flows away from a pollutant source.

It is still possible, despite these limitations, to observe some changes in pollutant concentrations registered by the stations over two time intervals. With the exception of dissolved solids, the data represent constituents tracked during two periods, 1974–1981 and 1980–1989. The data show that monitoring stations for four of the six pollutants tracked during both periods registered substantial reductions in pollutant concentrations. While the record for dissolved oxygen and nitrates also improved, the gains were lower during the 1980s than the 1970s.

Concentrations of fecal coliform and phosphorus show the sharpest declines. High concentrations of fecal coliform indicate a potential risk of infection to humans from pathogens associated with the bacteria. In both periods, the number of stations registering high concentrations sharply declined. USGS analysts attribute the drop to improved sewage treatment plants constructed under the Clean Water Act.

Phosphorus, a principle component of agricultural runoff, similarly declined during the 1980s. Stations showing decreases in phosphorus concentration outnumber stations showing increases by a ratio of 5 to 1. Declines are probably due to tighter controls on point sources and the phaseout of phosphorus in detergents.

Some substances also stabilized during the 1980s. Nitrate concentrations increased significantly during the 1970s, when the use of nitrogen-containing fertilizer peaked. Concentrations in stream water leveled off during the 1980s, with slightly more stations registering decreases than increases.

During the 1970s, dissolved oxygen concentrations sharply increased, indicating an improvement in stream-water quality. During the 1980s, increases were not as pronounced, yet monitoring stations that registered increases in dissolved oxygen outnumber stations that measured decreases. The gains occurred despite the fact that population increased 10 percent and real gross national product rose 30 percent during the same period. Given these factors, USGS researchers concluded that the trend in dissolved oxygen during the 1980s should be interpreted as generally positive, and may serve to show the effectiveness of the construction grants program and the industrial permitting system (Smith, Alexander, and Lanfear 1991, 116).

Regional Trends

The quality of water monitoring data tends to vary dramatically according to geographic location and scale. Here, we examine two closely monitored water bodies, the Great Lakes and the Chesapeake Bay, in order to illustrate the variation in environmental pressures and effects on different water bodies. In both areas, longstanding regional cooperative efforts to improve water quality have generated a comprehensive information base on pollutant loads, ambient concentrations, uptake, and ecological effects. The findings clearly cannot be generalized to national water quality trends. Still, they show how certain pollution problems respond to different policy strategies. Both regions have made significant progress in combating major pollution problems, but many challenges remain.

The Great Lakes. Twenty years ago, algal blooms, fish kills, and "dead zones"—areas of lakes where the oxygen supply was insufficient to support life—were common. According to the 1995 *State of the Great Lakes* report (MDEQ 1996), water quality has improved significantly. Still, pressures remain. Overall, the chief surface water challenges in the Great Lakes regions involve conventional pollutants such as phosphates and nitrates, as well as long-lived toxic pollutants. Conventional pollutants enter the Great Lakes primarily from agricultural runoff.

Phosphorus changes water chemistry and can lead to changes in water oxygen levels that eventually impact aquatic plants and the animals that feed off them. In response to these effects, the United States and Canada jointly established acceptable concentrations for phosphorus levels. The results are most prominent in Lake Erie, a lake once considered "dead." Phosphorus loadings in Lake Erie have declined from a high of more than 18,000 metric tons per year in the 1970s to about 11,000 in 1991, the last year for which data are available (U.S. CEQ 1996, 234). A significant portion of the decline is due to a ban on phosphate-containing laundry detergents. Declines in pollutant loads there have helped to improve water clarity and plant growth. Water quality improvement in Lake Erie also has helped to increase numbers of some fish. Rising fish stocks prompted the U.S./Canada Lake Erie Committee to raise the limit on allowable harvests of walleye in some parts of the lake from nine million fish in 1995 to eleven million in 1996 (U.S. CEQ 1996, 234). While pollution appears to be declining, other problems persist in the Great Lakes. For example, habitat loss and the introduction of foreign species such as the zebra mussel continue to degrade water quality and threaten native marine life.

Chesapeake Bay. The Chesapeake Bay is the nation's largest estuary. Historically, the bay has been a plentiful source of crabs, oysters, and fish. Stocks of each have experienced severe declines due to overharvesting and degraded water quality. Lack of dissolved oxygen is the bay's chief water quality problem. Oxygen depletion has increased in severity since colonial settlement as the watershed has been modified through deforestation, farming, and increases in population. Phosphorus and nitrogen are the major contributors to the process.

The decline of the bay became the focus of national attention during the 1980s. In an effort to halt the decline, the Chesapeake Bay Project, a federal-state partnership, was created in 1984. In 1987, the jurisdictions within the watershed established an ambitious set of voluntary measures to reduce pollutant loads into the bay by the year 2000. Participants agreed to reduce the quantity of phosphorus and nitrogen entering each of the bay's tributaries by 40 percent. Modeling of the origin and quantities of pollutants entering the bay indicates that the biggest success story in the bay involves phosphate. Since 1984, phosphate loads have declined by about 19 percent, due to upgrading of wastewater treatment plants and bans on phosphate detergents. Average monthly concentrations of total phosphorus declined significantly between 1985 and 1992. However, total nitrogen and dissolved oxygen showed little change over the period (U.S. EPA 1994d, 5).

While it is unlikely that the goals of the voluntary agreement will be met, water quality in some of the bay's major tributaries has improved since the passage of the Clean Water Act in 1972. During the 1960s and early 1970s, fish in the Potomac River asphyxiated in the summer. Due primarily to improvements in sewage treatment, the river again supports fish. Bald eagles and osprey that feed on the river's fish have returned as well. Remaining threats to the bay stem mostly from harder-to-solve problems such as automobile exhaust, runoff from city streets, and new residential and commercial development.

Coastal Uptake of Sewage. Population and development are among the greatest pressures on coastal areas. More than half of the U.S. population now lives in coastal regions. The trend is expected to continue through 2020 with growth in coastal states such as Texas, Florida, and, until recently, California (U.S. CEQ 1996, 248). Some coastal residents rely on beach waters not only for recreation but for their livelihood.

One indicator of overall contamination from sewage and other sources of pathogens is the National Shellfish Register compiled by the National Oceanic and Atmospheric Administration. Because shellfish filter a large volume of water, they reveal regional fecal coliform trends. Shellfish thus can indicate the effectiveness of sewage treatment systems, although fecal coliform also comes from nonpoint sources.

Bed closure trends reflect the effects of other variables besides actual shellfish bed quality. Tighter monitoring and improved reporting practices also determine how many shellfish beds are approved or prohibited for harvesting for human consumption. "Approved" status means that shellfish may be harvested at all times. "Prohibited" means that the catch is unfit for human consumption.

Between 1966 and 1990, shellfish bed closure rates declined in the Pacific and Southeast (see Figure 5). However, in the same period the number of prohibited acres rose dramatically in the Gulf of Mexico and increased slightly in the Northeast. Half of all shellfish growing waters in the Gulf of Mexico region fail to meet criteria for approved waters. Gulf pollution is attributed in part to drainage from the large gulf watershed. Population in the gulf watershed has more than doubled since 1960, accounting for much of the pollution pressure.

Between 1985 and 1990, the last period for which data are available, the number of approved shellfish-growing waters fell nationwide by 6 percent. The drop was associated with a 1.2 million acre increase in waters prohibited for shellfishing. NOAA attributes the recent drop to increased coastal development, which tends to increase urban runoff, as well as the possibility of sewage leaks from faulty septic systems or marina development zones. Fecal coliform readings were highest in the nation's most productive estuaries, including Chesapeake Bay and Puget Sound (U.S. DOC 1991).

Human Health Effects

While no systematic data exist to examine the effects of water pollution on people and other living things, some studies suggest that despite declines in effluents, swimming, fishing, or drinking from many water bodies in the United States remains unsafe. Chemicals, bacteria, and viruses are water contaminants that can make people sick. Exposure can occur through numerous routes, including contact through swimming or by drinking water. The processes by which humans are exposed as well as

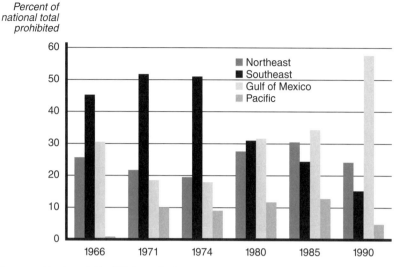

Figure 5. Regional Shellfish Bed Closures as Percentage of National Total, 1966–1990.

Source: U.S. CEQ 1990, 309–10.

the exact ways that contaminants make people sick are still not well understood. The effects of swimming in polluted water are among the least documented, primarily because beach-related illness tends to be underreported to medical personnel.

Constance, Sullivan, and Barron (1989) found that Los Angeles County lifeguards who worked along some of the county's most severely polluted waters had a higher than normal rate of illness based on worker's compensation claims. Critics of the study argue that other factors such as lifestyle could account for illness among lifeguards.

Another measure of coastal water quality is the number of beach closures due to contamination. The Natural Resources Defense Council has charted state-reported beach closures and advisories from 1988 to 1995 (NRDC 1996). While NRDC reports do not document actual illness, the designations strongly suggest that swimming is unsafe in some coastal areas. According to NRDC, in 1995 nearly 4,000 beaches were closed or under advisories issued by the states for at least one day. Primarily, states closed beaches or issued advisories due to sewage overflows, stormwater runoff, and sewage treatment plant malfunctions.

Evidence suggests that drinking water supplies still are not always fit for human consumption. In recent years, wide-scale illness associated with contaminated drinking water has placed the issue in the public spotlight. The most notorious episode occurred in 1993, when 183,000 residents in Milwaukee, Wisconsin became ill after drinking public water contaminated by the protozoa cryptosporidia, an intestinal parasite. The episode contributed to the deaths of 100 people. More recently, residents and workers in the nation's capital were warned repeatedly not to drink from the tap during the summer of 1996 because monitoring on several occasions showed unsafe concentrations of contaminants.

Unfortunately, no data sources exist to document long-term national drinking water quality trends. The Centers for Disease Control (CDC) report outbreaks of waterborne disease, but the outbreaks are not identified by year. Between 1972 and 1990, CDC recorded a total of 554 disease outbreaks, affecting 136,000 people, that were linked to public water supplies (Levine, Stephenson, and Craun 1990). Between 1971 and 1985, more disease outbreaks were reported than in any previous fifteen-year period since 1920. Outbreaks were highest during the 1970s and dropped off between 1984 and 1990.

The only other national perspective comes from data on violations recorded by EPA's drinking water program. The agency records compliance rates among states and territories that have received authority under the Safe Drinking Water Act (SDWA) to administer and supervise their public water systems. In order to receive such authority, localities must demonstrate to EPA that they have adopted drinking water regulations that are at least as stringent as federal requirements. In FY1995, regulations under the law were in effect for 81 individual contaminants, including organic and inorganic chemicals, microbiological contaminants, and radionuclides. The SDWA sets contaminant levels to insure that water is fit for human consumption. Each year, EPA surveys localities to determine the incidence of violations. Most reports reflect violations in community water systems (CWS). CWS supply water year-round to the same population, usually people in residential areas. In 1995, 55,633 such systems supplied water to nearly 244 million people. Other drinking water systems include public water systems (such as school systems) that maintain their own water supplies and those that cater to customers in nonresidential areas, such as campgrounds and motels. Since CWS represent the majority of drinking water systems, only those rates are shown here.

Typically, municipalities monitor unfiltered supplies and, when necessary, treat water to attain drinking water standards. Compliance rates are based on whether water remained within federal maximum contaminant levels (MCLs) for criteria such as microbial concentration, turbidity, and inorganic and organic chemicals. For some contaminants, the SDWA requires treatment techniques in lieu of MCLs; since 1993, water systems could also be found out of compliance for improper treatment techniques. Finally, systems may also be in violation for monitoring and reporting failures.

EPA reports that violations declined between 1986 and 1988, increased sharply in 1992, and decreased in 1993. (See Figure 6.) EPA attributes the increase to new reporting requirements and stiffer regulations. Over the period, the number of systems in violation ranged between 18,000 and 20,000 (U.S. EPA 1995b, 22). Preliminary data for fiscal years 1994 and 1995 show an increase in the number of monitoring and reporting violations (U.S. EPA 1996f, 10).

In addition to monitoring drinking water systems, EPA is trying to help states and localities improve the quality of groundwater, which provides drinking water for 51 percent of the total population (U.S. EPA 1994b, 71; 1995d, 124–25). Unfortunately, no complete or accurate profile of national groundwater quality exists. The deficiency is due to several factors, including the expense of generating groundwater monitoring data, variation in size and configuration among aquifers across the United States, and variations among states in monitoring and reporting techniques. To correct the deficiency, EPA has begun to work with states to develop guidelines to evaluate groundwater quality (U.S. EPA 1995d, 124–25).

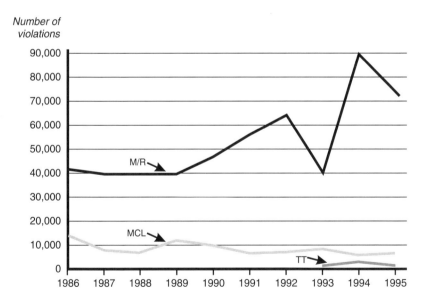

Figure 6. Number and Type of Violations in Community Water Systems, FY1986–1995.

Notes: M/R denotes monitoring and reporting violations; MCL is violations in Maximum Contaminant Levels; TT indicates inadequate treatment techniques.

Sources: U.S. EPA 1995b, 22; 1996f, 10.

MUNICIPAL SOLID WASTE

In contrast to its role in regulating air and water pollution, the federal government plays a limited role in municipal solid waste management. While EPA banned ocean dumping of solid waste, states, regional authorities, and municipalities implement municipal solid waste programs. EPA defines municipal solid waste as a subset of nonhazardous solid waste. Municipal solid waste excludes construction and demolition waste, industrial process waste, and other wastes that go to a landfill. EPA provides guidance to states and localities on how to manage wastes.

To guide local management efforts, EPA provides estimates of waste generation trends. National municipal solid waste data illustrate both long-term trends in waste generation and composition of the waste stream in the United States. The data show waste generation patterns from 1960 to 1994 and include projections to the year 2010 (U.S. EPA 1996b).

Like the air emissions estimates, data on municipal solid waste are based on engineering calculations and not on real garbage samples. The data therefore fail to reflect local variations that arise from differences in climate and commercial activity. For example, lawn waste varies by season in some places and remains steady in others. Commercial activity also changes the composition of the waste stream.

In addition to omitting regional variation, the characterization fails to account for some products in the waste stream. The reports do not reflect residues such as jelly left in a jar or pesticides in an aerosol can. The characterization also applies only to products for which data have been developed. Data series for substances such as pigments and inks associated with plastics have not yet been developed.

Despite their limitations, the data show some encouraging trends. Solid waste management strategies such as recycling and composting have recently helped to reduce generation rates. However, the long-term record is less encouraging. Municipal solid waste generation more than doubled in three decades, outstripping increases in population growth. (See Figure 7.) Waste generation grew from 87 million tons in 1960 to 209 million tons in 1994. Per capita generation climbed from 2.7 pounds per person per year to 4.4 over the same period. EPA reports that economic growth is primarily responsible for the increases (U.S. EPA 1996b, 118).

HAZARDOUS WASTE

Hazardous waste is regulated under two laws, the Resource Conservation and Recovery Act (RCRA) of 1976, as amended, and the Comprehensive Environmental Response, Compensation, and Liability Act (CER-

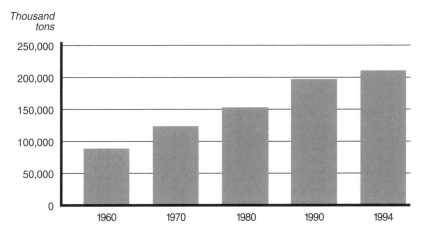

Figure 7. Total Municipal Solid Waste Generated, 1960–1994.

Source: U.S. EPA 1996b, Table 18 at 67, 118.

CLA). RCRA Section C regulates the handling, storage, treatment, and disposal of hazardous waste, while CERCLA (or Superfund) provides for cleanup of inactive or abandoned sites.

In contrast to "solid" waste or "toxic" substances, the term "hazardous" waste refers to solids or liquids or combinations thereof that may "cause or contribute to an increase in mortality or an increase in serious irreversible or incapacitating reversible illness" or any wastes that may "pose a substantial threat to human health" when improperly handled (42 U.S.C. section 1004 (5) 6903). Given the potential effects of such substances on humans and the natural environment, disappointingly few data exist to illustrate the long-term nationwide trends in hazardous waste generation (U.S. GAO 1991).

To help track hazardous wastes regulated under RCRA, EPA in cooperation with states biennially collects information regarding the generation, management, and final disposition of hazardous wastes. While data may eventually be collected and reported in a manner that provides insights into national hazardous waste generation trends, to date they appear to suffer from many of the same weaknesses as the 305(b) reporting provisions under the Clean Water Act. Like the water quality surveys, the biennial reports are based on surveys administered by states. Data cannot be aggregated into a comprehensive national portrait because state procedures for gathering and reporting the information vary. For example, in some states, administrators are unable to track down and report on the total number of hazardous waste generators. In other cases, state mailing lists used to identify survey respondents are outdated and thus fail to identify new generators (Gottlieb 1995, 136). Other data problems owe more to RCRA's often confusing definitions of what constitutes hazardous waste. As a result, reporting forms tend to contain errors. One study found that in some cases biennial reporting forms were left blank or not reported at all (U.S. GAO 1991).

To date, states have submitted biennial reports from 1985 to 1993. (See Figure 8.) Due primarily to disparities in state reporting systems, uniform data elements on a nationwide basis only became available with the 1989 reports. But even for 1989, many common data elements are missing from state reports. Missing data elements include generator identity (standard industrial classification code) and waste description, as well as data on waste minimization (Gottlieb 1995). Data for 1989 and 1991 are not directly comparable because EPA in 1991 expanded the waste code, nearly doubling the number of wastes to be reported. Data for 1991 and 1993, the most recent years for which reports are available, reflect the expanded waste code categories and thus may be compared. However, two years are insufficient to present an accurate picture of national trends.

Perhaps the greatest shortcoming of the biennial reports is that they only reflect a subset of all hazardous waste generated in the United

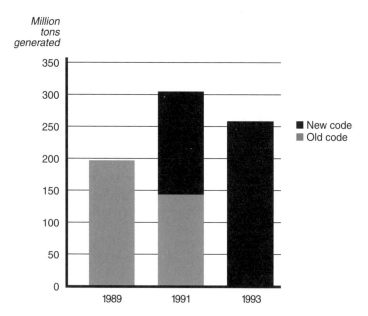

Figure 8. Hazardous Waste Reported under RCRA, 1989–1993.

Note: "Code" refers to hazardous waste codes added under the expansion of RCRA reporting categories.

Source: U.S. EPA 1996e, executive summary.

States. When the exemptions under RCRA are factored in, hazardous waste generation is much larger than the biennial reports suggest. The Department of Energy estimates industrial amounts of hazardous waste to be on the order of 750 million tons—more than double the amount reported under RCRA in 1991 (U.S. DOE 1991). Several sectors (such as mining, petroleum production, electric power generation) as well as small sources are granted exclusions under hazardous waste provisions of RCRA. The biennial reports also omit hazardous wastes reported under air or water statutes.

While the overall amounts reported thus understate hazardous waste generation, EPA's reporting categories may overstate wastes for certain sectors, such as the chemical industry (Allen and Jain 1992). The biennial reports require firms to estimate "waste streams," which often consist of water, soil, and other nonhazardous materials. Under RCRA, however, the entire stream, including disposal medium, is categorized as hazardous. As a result, wastes can appear enormous yet contain very few toxic or hazardous substances.

In order to obtain a better estimate of waste, Allen and Jain analyzed production data for the chemical industry, which accounts for 80 to 90 percent of hazardous waste generation reported under RCRA. Comparing raw materials to final products to determine product yields, they assumed that what was not going into the final product was either waste or byproduct. Overall, they found that dry hazardous wastes only represent a small percentage of materials released to waste streams (Allen and Jain 1992).

Despite current shortcomings, biennial survey data collected by EPA under RCRA data may eventually supply a better long-term picture of generation trends. In early 1997, EPA completed the first phase of the Waste Information Initiative, an effort to improve the match between information needed to improve waste management and the type of information industry is required by EPA to report (U.S. EPA 1997a). Biennial reports show that in 1993, 24,362 large quantity generators produced 258 million tons of hazardous wastes regulated by RCRA. The 1993 data show an increase of 936 generators and a decrease of approximately 47 million tons of waste compared with 1991 (U.S. EPA 1996e). In 1991, the RCRA reporting categories under the toxicity characteristic rule added twenty-five new hazardous waste codes and as much as 167 million tons of new hazardous waste.

Contaminated Sites

Superfund extends to all types of hazardous waste sites, including former municipal landfills, co-disposal landfills, waste handling and disposal facilities, recycling facilities, and industrial facilities (Probst and others

1995, 35–36). Due to the multiple types of wastes and multiple parties potentially responsible for contamination, the pace of cleanup of abandoned or inactive hazardous waste sites has been characterized as slow. Cleanups typically consist of short-term "emergency" removal actions to control or remove substances that pose the most immediate dangers, followed by a lengthy remedial phase that typically involves some type of construction or excavation at the site.

As of 1995, EPA had taken approximately 3,042 actions to remove the most hazardous substances at the nation's high priority sites (U.S. EPA 1996g). Permanent cleanup of Superfund sites can take many years, because negotiating the remedies to be applied and who should pay for them can be a rancorous process; also, the actual cleanup, especially of contaminated groundwater, is challenging. As of June 1996, construction work had been completed at 410 of the 1,227 sites on the Superfund National Priorities List (U.S. EPA 1996g).

TOXICS

Toxics are substances that pose serious risks to human health and the environment. The category is somewhat ambiguous in part because it is difficult to determine with certainty how substances affect human and environmental health and in part because the effect usually depends on the amount of the substance to which a person is exposed. Comprised primarily of organic chemicals and metals, toxics are generally understood to be separate from conventional pollutants such as solid waste, silt, or bacteria.

The manufacture and use of toxic substances in the United States exploded after World War II. Before then, production of synthetic organic chemicals totaled fewer than one billion pounds a year. By 1976, production had increased to 169.2 billion pounds per year. That year, in response to problems posed by substances such as PCBs, asbestos, lead, and mercury, Congress passed the Toxic Substances Control Act (TSCA). Drafters believed that the statute would help avert future chemical problems (Gottlieb 1995).

TSCA gives EPA the authority to screen, test, and, when necessary, ban substances that pose "unreasonable risk" to human health and the environment. TSCA also authorizes EPA, based on notices of intent to manufacture new chemicals, to ban or restrict substances *before* they become commercially available and thus embedded in the nation's economic and social fabric. TSCA requires EPA to balance risk reduction strategies against economic costs.

Since the statute is not aimed at controlling chemical residuals, TSCA contains no monitoring provisions for toxic substances. Instead, EPA compiles different inventories of toxic substances under various statutes. For example, the Clean Air Act lists nearly 200 substances as "toxic." The Toxics Release Inventory contains over 600 substances on which manufacturers must report. However, no single, comprehensive data source exists for toxics. Examining production data on select organic and inorganic chemicals is one way to gauge whether pollution levels are declining.

While we cannot definitively link output trends to toxic provisions under the various statutes, it is nonetheless interesting to examine whether production volumes of some toxic substances correspond to the number of laws and ensuing standards and monitoring efforts directed at them. For example, the previous chapters illustrate declines in both ambient lead and lead emissions to air and water. One would expect lead production to show similar trends.

To help narrow the list of the more than 70,000 potential chemicals to study, we tracked, from 1970 onward, production data for 17 substances (see Table 6) targeted under a voluntary EPA toxic reduction program known as "33/50." EPA selected the 17 program chemicals primarily because they pose the most serious environmental and health concerns and also are produced in high volume. The program takes its name from EPA's goal to encourage firms to voluntarily reduce the substances 33 percent by 1992 and 50 percent by 1995 (U.S. EPA 1995e, 274).

Historically, annual production data for organic and inorganic substances have been available through the International Trade Commission (ITC) and the Department of Interior's former Bureau of Mines. Due to federal budget cutbacks, both data sources were discontinued in 1996.

We present production data from ITC and the Bureau of Mines through 1993, the last year for which we obtained a uniform set of data from the Bureau of Mines, and contrast these data with total toxic chemi-

Table 6. Target TRI Chemicals.

Organic chemicals	Inorganic chemicals
Benzene	Cadmium and cadmium compounds
Carbon tetrachloride	Chromium and chromium compounds
Chloroform	Cyanide compounds
Dichloromethane	Lead and lead compounds
Methyl ethyl ketone	Mercury and mercury compounds
Methyl isobutyl ketone	Nickel and nickel compounds
Tetrachloroethylene	
Toluene	
1,1,1-Trichloroethane	
Trichloroethylene	
Xylenes	

cal release and transfer information reported by manufacturers to EPA for that year. Data on metals and inorganic chemicals are primarily compiled for economic analysis and are not intended to reflect the state of the environment or exposure risk.

Because production data are not used to track environmental impacts, ITC and the Bureau of Mines collect, verify, and report the data differently than environmental statistics complied by EPA and other agencies that monitor environmental progress. For example, EPA tracks four substances under the xylene group, while ITC tracks only two. As a result, production figures should not be directly compared with environmental monitoring data compiled by EPA.

Finally, it should be noted that the figures often only capture domestic production. While chemicals imported into the United States are regulated under TSCA, imports are not counted under the U.S. production and sales data supplied by the International Trade Commission. The Bureau of Mines accounts for imported materials as consumption in the United States. Our figures distinguish between production and consumption.

Organic Chemicals

For the most part, the twenty-four-year record for the organic chemicals shows no discernible trend. (See Table 7.) Some substances decreased or increased slightly; only a few showed marked increases. These latter include benzene, trichloroethane, and the xylene group. In contrast, production of carbon tetrachloride, an ozone depleter, declined.

Associated with petroleum production, xylenes, benzene, and toluene are related in both origin and function. In all likelihood, production increases occurred because the substances are vital chemical intermediates in many consumer products. Xylene production increased 177 percent from 1970 to 1993. Xylenes contribute to ozone formation in the lower atmosphere. Humans exposed to high doses of xylene can suffer from dizziness, headaches, and possibly even liver damage.

Benzene, a suspected carcinogen, is easier to produce than xylenes, but standards for benzene are much more stringent than for xylenes. It is possible that standards encouraged producers to manufacture xylenes instead of benzene (Roque 1995). Benzene production nevertheless increased 50 percent over the period.

Inorganic Chemicals

Unlike organic chemicals, the metals listed under 33/50 do not degrade and are not destroyed, although some may be converted to a less toxic or more toxic form. For example, some facilities reduce hexavalent chromium

Table 7. Selected U.S. Commodity Chemical Production or Consumption, 1970–1993 (million pounds, unless noted).

Chemical	1970	1975	1980	1985	1990	1993	% change, 1970–93
Benzene (mgal)	1,134	1,024	1,534	1,283	1,699	1,667	47
Carbon tetrachloride	1,011	906	710	646	413	na	na
Chloroform	na	na	273	275	484	476	na
Methyl ethyl ketone	480	425	587	537	465	539	12
Trichloroethane (methylchloroform)	366	459	692	869	803	452	23
Methylene chloride	402	497	564	467	461	354	–12
Tetrachloroethylene (perchloroethylene)	707	679	765	678	372	271	–62
Toluene (mgal)	830	705	1,010	698	861	880	6
Xylenes (o-xylene, p-xylene)	2,389	3,187	5,233	5,454	6,143	6,622	177
Trichloroethylene	.61	.29	.27	na	na	na	
Methyl isobutyl ketone	.19	.15	.17	.13	na	na	
Cadmium production	9.47	4.39	3.48	3.53	3.70	2.41	–75
Chromium consumption	858	512	505.0	585.5	985.6	na	
Nickel consumption	311.4	293	312.6	324.0	350.6	na	
Lead production	1,334	1,272	1,120	1,089	682	660	–45
Mercury consumption	4.67	3.87	4.48	3.79	1.59	1.23	–74
Cyanide compounds	na	na	na	na	na	na	na

Sources: U.S. ITC 1993; U.S. DOI 1992, 331–34, 886; Woodbury, Edelstein, and Jasinski 1993, Table 1; Llewellyn 1994, 17; Papp 1994, 136; Jasinski 1994, 11; C&EN 1985, 27; C&EN 1990, 38; C&EN 1996, 42.

Notes: Benzene and toluene are reported in million gallons; benzene data exclude tar distillers and coke oven operators; chloroform statistics do not appear for earlier years; data are not reported when substance is produced by three or fewer than three manufacturers. na = Data not available.

(a suspected carcinogen) into trivalent chromium before releasing or transferring it offsite. Because they do not degrade, metals accumulate and persist in the natural environment.

Overall, the metals record is mixed, with a few pronounced trends. Production and domestic consumption of cadmium and lead dropped significantly. Hexavalent chromium, which is released to air in the chromium plating process, increased along with nickel consumption. Both also are inputs into stainless steel. Hexavalent chromium is associated with carcinogenic, mutagenic, and teratogenic health effects, and also has been linked to adverse environmental effects. Because the principal exposure route to humans is through inhalation during processing, the toxic metal is regulated primarily under occupational safety laws. Cadmium production started to decline in 1970 and has dropped 75 percent since then (Llewellyn 1994, 13).

Lead is not a known carcinogen, but exposure in high doses has both acute and chronic effects. Acute (that is, short-term) exposure symptoms include central nervous system disorders. Chronic effects from exposure to lead-based paint or contaminated soils include developmental effects such as severe mental impairment. While lead emissions to air dropped nearly 90 percent since 1970 through the gasoline phaseout and product bans, primary production only dropped 45 percent. The disparity may exist because lead is still used in batteries and as a radiation shield in medical equipment (Woodbury, Edelstein, and Jasinski 1993, Table 1).

Mercury production peaked at 6.6 million pounds in 1964, before most mercury regulations existed. Mercury is toxic in humans and accumulates in fish tissue. The 1960s witnessed several severe outbreaks of mercury poisoning in humans. During that time, high mercury concentrations were reported in Great Lakes fish as well. The decrease resulted primarily from a manufacturing process change that substituted other substances for mercury (Jasinski 1994, 11).

Laws probably account for continued mercury declines. The toxic substance is now heavily regulated under a number of different environmental statutes, including the Clean Air Act; Clean Water Act; Safe Drinking Water Act; and Federal Insecticide, Rodenticide, and Fungicide Act, which banned the use of mercury in pesticides.

Emissions

Although no systematic data exist to track toxic emissions into the environment, the Toxics Release Inventory (TRI) provides a useful picture of release and transfer trends for individual chemicals and individual facilities. (See Figure 9.) Data for any individual year also supplement individual air, water, and waste reporting provisions by requiring firms to list releases and transfers by medium. To date, however, the TRI is not adequate to track trends in long-term toxic chemical emissions.

The TRI requires manufacturers in the United States to report annually to EPA on releases and offsite transfers of certain toxic chemicals. Established in 1987 through the Superfund Reauthorization Amendments Title 313, the TRI aims to help communities avert chemical accidents and to track certain types of chemicals released and transferred by manufacturers. Manufacturers employing more than ten workers and producing more than 25,000 pounds of the TRI list's substances, or firms that use more than 10,000 pounds of these substances per year, are required to report annual releases and transfers of approximately 350 chemicals for the 1987 through

Billion pounds

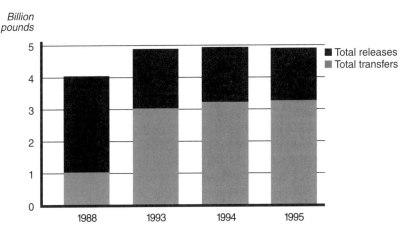

Figure 9. TRI Releases and Transfers, 1988, 1993–1995.

Note: Data used for year-to-year comparison differ from 1995 reporting year totals; 1988 transfer data are not comparable with other reporting years.

Source: U.S. EPA 1997b, 118.

1994 reporting years, expanded to 643 in 1994 (U.S. EPA 1997b, 21). Manufacturing facilities are defined as facilities in Standard Industrial Classification primary codes 20–39; these codes include, among others, chemicals, petroleum refining, primary metals, fabricated metals, paper, plastics, and transportation equipment. In 1997, EPA expanded TRI reporting requirements to include seven additional industrial sectors, including electricity generating plants and metal and coal mining facilities (U.S. EPA 1997b). Since 1987, the first reporting year, EPA has deleted some chemicals from the list and changed reporting requirements for others. For example, in an effort to better monitor whether firms are attempting to reduce toxics, Congress added two new reporting categories to the TRI effective as of 1991. The categories require firms to report on transfers to recycling and energy recovery. Because of the new categories, data from 1991 are not directly comparable with earlier years. To aid in comparison, EPA developed a control list of chemicals to ensure that fluctuations are not due to changes in reporting methods. Finally, the TRI was expanded in 1994 by 286 chemicals and new chemical categories. As a result, year-to-year comparisons for TRI chemicals may differ slightly from totals reported for 1995, the most recent year for which data on releases are available.

In 1995, reporting facilities released 2.2 billion pounds of TRI-listed chemicals into the environment (U.S. EPA 1997b, Table 4-1) and transferred about 3.5 billion pounds of TRI chemicals offsite (U.S. EPA 1997b, Table 4-2). Air emissions constituted nearly 71 percent of all toxic chemical releases in 1995 (U.S. EPA 1997b, Figure 4-1). Air releases are followed by on-site land disposal and injection of chemicals into underground wells. In 1995, about 13 percent of all TRI chemicals were disposed of on land, while 11 percent were injected into underground wells. Releases to surface waters in 1995 comprised the smallest percentage of all TRI categories, about 6 percent. Manufacturers released about 136 million pounds of toxics directly into lakes, coastal areas, and other water bodies. About 240 million pounds of toxic chemicals were sent to public sewage treatment facilities in 1995 (U.S. EPA 1997b, 19). The good news from the TRI data is that a dramatic reduction has taken place in direct releases to the environment. The bad news is that the total amount of waste generated has increased. This last result is not surprising, given the huge expansion in the U.S. economy during the period covered.

The TRI clearly portrays releases and transfers for some industrial sectors. When production and consumption data for organic and inorganic materials are contrasted with emissions and transfers reported to the TRI, it is clear that the inventory represents just a small subset of materials that may flow into the environment. When expressed as a percentage of production, the inventory at best accounts for less than 50 percent of all but one of the 33/50 substances. (See Table 8.) Emissions and transfer data reported to TRI reflect less than 1 percent of the total production volume of toxic substances such as benzene. The inventory accounts for about 45 percent of lead released and transferred in the United States. In the case of cadmium, which is mostly imported, emissions and transfers appear to greatly exceed U.S. production.

The TRI numbers fail to match production figures for several reasons. As mentioned, many inorganic substances are now imported into the United States. In addition, only a subset of all potential chemical users is required to file TRI reports. Presently, potential emissions sources such as service and mining industries are exempted. Another major reason that the numbers may fail to match is because not all toxics used are released or transferred: some toxic materials may be stored on-site or contained in intermediate and final products and thus not reported to the TRI.

Exposure

Some toxic substances persist in the environment and in people for long periods of time. Data on production, releases, and transfers help us to understand whether overall quantities released to the environment may be increasing. The data may also provide some indication of just how much material has been released over the past few decades. However, some of the most important monitoring data are those that show concentrations of toxic substances in animals and in humans. Several data sources track individual substances over long periods.

Table 8. TRI Releases and Transfers as a Percentage of Production, 1993 (million pounds, unless noted).

	Production	TRI release and transfer	TRI as percentage of production
Benzene (mgal)	1,677	15.8	0.9
Carbon tetrachloride	na	3.4	na
Chloroform	476	17.4	3.7
Methyl ethyl ketone	556	164	29.5
Trichloroethane	452	85.7	19
Methylene chloride	354	101	28.5
Tetrachlroethylene	271	21.5	7.9
Toluene (mgal)	833	318	38
Xylenes	6,641	240	3.6
Trichloroethylene	na	40.6	na
Methyl isobutyl ketone	na	62.6	na
Cadmium (production)	2	5.7	300
Chromium (consumption)	na	178	na
Lead (production)	737	335	45
Mercury (consumption)	1.2	0.1	8
Nickel (consumption)	na	109.7	na

Sources: U.S. ITC 1993; U.S. DOI 1992; U.S. EPA 1996f, 284–87.

Note: Cadmium figure is probably lower than TRI because it does not reflect imports of the substance into the United States. na = Data not available.

One of these sources is provided by the U.S. Geological Survey in water quality reports. In addition to tracking concentrations of some conventional pollutants in some streams, the USGS tracked concentrations of several metals (lead, zinc, chromium, silver, and arsenic) during the 1970s. The agency now uses biological indicators, including concentrations of metals in fish tissue samples. Pronounced trends were apparent during the 1970s. Only a few streams showed increased concentrations of these metals. Many more showed fewer metals. Lead reductions were among the most pronounced (Smith, Alexander, and Wollman 1987, 18).

A separate monitoring system maintained by NOAA measures sediment and fish and shellfish contamination. The National Status and Trends program tracks only a handful of pollutants, and many NOAA national monitoring stations were deliberately located away from heavily industrialized areas in order to provide a broader gauge of ambient conditions. NOAA national trends data are generally consistent with the stream water quality trends. Overall, concentrations of toxic metals are constant or declining in coastal waters. PCBs showed the strongest overall declines. Copper was the only metal to show increasing trends at all sites (O'Connor 1991).

While emissions and concentrations in air or water may be declining, many toxics persist in hard-to-remove sediments. Such "persistent" materials may concentrate in fish tissue. Exposure to toxic substances is highest for people who eat fish from regions with high toxic sediment concentrations. Possible human health effects related to fish consumption remain unclear.

The only definitive links between exposure from fish consumption and health effects in wildlife is for birds exposed to the pesticide DDT (Conservation Foundation 1990, 133). One national level data source examines changes over time in some trace elements and organic compounds in whole finfish (Schmitt and Brumbaugh 1990). Compiled by the Fish and Wildlife Service, the data show trends in toxic concentrations in the whole fish, not simply the parts people consume.

Fish samples collected in major rivers in the United States indicate that arsenic, cadmium, and lead decreased 50 to 63 percent between 1976 and 1986. Mercury concentrations remained constant. Fish-tissue concentrations of PCBs decreased by more than 60 percent between 1970 and 1986. The data are generally consistent with those developed by the Geological Survey as well as for the Great Lakes, where toxic chemicals are a lingering reminder of the region's heavy industrial history.

Research conducted jointly by the United States and Canada indicates that concentrations of some organic toxic chemicals are dropping (U.S. CEQ 1995, 491). Fish tissue samples show that PCBs have declined nearly 90 percent since the mid-1970s. Lake Michigan has the longest monitoring history and thus best illustrates the trend. (See Figure 10.)

PCBs in wildlife have declined dramatically since the Toxic Substances Control Act banned most uses of the substance in 1976. However, the PCB concentrations in fish persist well above levels set to protect public health. The persistent PCB burdens in some fish, mammals, and birds still may impair reproductive success. While PCB concentrations declined, they remain at about 180 times the target goal of 0.014 parts per million. Body burdens of PCBs in a colony of Forster's terns near Green Bay, Wisconsin declined by 66 percent while hatching success tripled between 1983 and 1988. However, the terns' offspring continued to suffer "wasting" and other fatal health problems that may have resulted from the lingering effects of PCBs in adult birds (Conservation Foundation 1990, 135).

Some of the most comprehensive data on toxic concentrations in humans derives from blood lead samples. Screening programs conducted by the Centers for Disease Control and Prevention have produced data samples fairly representative of the population in the United States. Due primarily to the enormous quantities of lead released during the era of

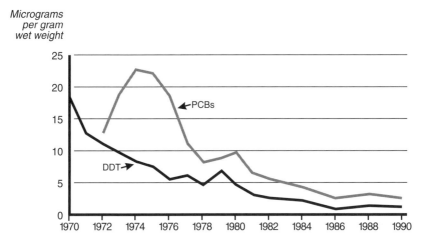

Figure 10. PCB and DDT Contaminant Levels in Lake Michigan Trout, 1970–1990.

Source: U.S. CEQ 1995, 491.

leaded fuel use, researchers estimate that lead body burdens in North Americans are as much as 300 times those of Native Americans who lived before Europeans arrived. The high levels show that while lead emissions have declined steeply, most of the more than 300 million tons of lead ever mined remains in the environment (NRC 1993, 18). Despite the high incidence of lead exposure and background lead in the environment, the average level of lead found in the blood of humans has declined 78 percent from 1976–1980 to 1988–1991 (Dodell 1994).

PESTICIDES

Pesticides have expanded agricultural productivity and provided U.S. citizens with one of the most abundant food supplies in human history, but they also can result in adverse health and environmental effects. EPA regulates pesticides under two statutes: the Federal Insecticide, Rodenticide, and Fungicide Act and the Federal Food, Drug and Cosmetic Act.

EPA defines pesticides as "any agent used to kill or control undesired insects, weeds, rodents, fungi, bacteria or other organicides" (U.S. EPA 1991, 2). The term therefore includes insecticides, nematocides, and acaracides, as well as disinfectants, fumigants, and plant growth regulators. "Conventional pesticides" consist of herbicides, insecticides, and fungicides. Approximately three-quarters of conventional pesticides are used in agriculture; the remainder is split roughly evenly between house-hold and commercial uses (U.S. EPA 1991a, 2–3).

In addition to registering new and existing pesticides that farmers apply to fields, EPA sets tolerances (the maximum amount of a residue that is permitted in or on a food). The federal Food and Drug Administration (FDA) is charged with enforcing tolerances in domestic and imported foods. The U.S. Department of Agriculture (USDA) monitors application rates of pesticides on farm crops.

As the data illustrate, pesticide use on select crops doubled between 1964 and 1982, when use peaked at the height of the farming boom. (See Table 9.) Since then, land area planted with crops monitored for pesticide use has decreased (USDA 1994, 86). Among the different types of pesticides used, the most substantial decline is for insecticides. During the 1960s, insecticides accounted for half of all pesticides used on crops.

Insecticide use has dropped from 136 million pounds in 1976 to under 60 million pounds in recent years. The declines are mostly tied to cotton. Insecticides commonly used on this fiber, such as DDT and toxaphene, have been banned, and cotton growers also are able to use fewer insecticides due to successful efforts to eradicate pests such as the destructive boll weevil. Development of integrated pest management techniques has also reduced the need for pesticides.

Declines in insecticide use have largely been offset by increased herbicide use. Herbicides increased from 55 million pounds in 1964 to 465 million pounds in 1982, accounting for three-quarters of all pesticides used. However, herbicide use has declined somewhat since then, to 388 million pounds in 1992 (USDA 1994, 86).

Ambient Concentrations

Application rates give some indication of what pesticides may be going into the environment, but they fail to indicate where these substances are ultimately deposited. While concentrations located in some water bodies have declined, irrigation and harvest continue to cause pesticides to migrate into water supplies and to linger on food crops. Smith, Alexander, and Lanfear (1991, 134) reference other USGS studies that found up to half of all surface water in a ten-state area of the Midwest contains concentrations of pesticides that exceed EPA's maximum contaminant levels and lifetime health-advisory levels.

Repeated samples from surface water revealed detectable concentrations of several herbicides. Atrazine, one of the most commonly used herbicides for weed control, occurred year-round in a majority (52 percent) of

Table 9. Pesticide Use on Selected U.S. Crops by Major Type, 1964–1992.

1,000 lbs active ingredient	1964	1966	1971	1976	1982	1990	1991	1992	Percentage change 1964–92
Herbicides	54,884	87,351	198,949	368,422	464,596	376,367	368,268	388,175	607
Insecticides	128,167	121,717	137,808	135,920	84,793	56,621	51,053	57,855	–55
Fungicides	21,715	21,660	30,906	29,546	27,519	31,641	33,112	37,358	72
Other pesticides	27,983	24,233	31,565	31,072	35,417	68,971	80,893	90,519	223
Total	232,750	254,961	399,228	564,960	612,325	533,600	533,346	573,907	147

Source: USDA 1994, 86.

the streams sampled in the midwestern study area. USGS researchers also detected concentrations of alachlor and two other herbicides, cyanazine and simazine, in up to 49 percent of the streams sampled.

In addition to washing off fields, pesticide residues remain in harvested food. According to the FDA's pesticide residue monitoring program, levels of pesticide residues in the domestic food supply are generally well below established limits. The FDA samples three categories of domestically produced and imported foods: meat and shellfish, grains, and dairy. Long-term residue trends are encouraging. Sampling data since 1978 indicate that the biggest residue declines have occurred in dairy products (86 percent). These declines were followed by a 35 percent decline in residues on meat and shellfish. The data show no discernible trend for grain products (U.S. FDA 1996, 8–9).

The 1996 figures for 5,474 food samples grown in the United States categorize residues as containing either no residue, nonviolative residue, or violative residue. Violative is defined as "a residue that exceeds a tolerance or a residue at a level of regulatory significance for which no tolerance has been established in the sampled food." Of all domestic food sampled in 1994, 99 percent contained no violative residues. Less than 1 percent of the fruit and 2 percent of the vegetable samples contained pesticide residues that exceeded safety standards. About 44 percent of all sampled fruits and 66 percent of vegetables sampled contained no residue. In the dairy/milk/egg category, about 93 percent of the samples were found to contain no residue.

The FDA also tests some imported food to determine whether unregistered pesticides sold abroad enter the United States as residues. In 1994, FDA targeted specific unregistered pesticides based on foreign usage. A total of 5,448 samples of imported foods from twenty-five countries was analyzed. Less than 1 percent of the sample contained residues that exceeded tolerance limits (U.S. FDA 1996, 10).

Pesticides can flow from farm fields into wells and water bodies, where they collect in plants, sediments, and fish. Another indicator of pesticide presence in stream water comes from finfish data collected by the Fish and Wildlife Service (FWS). In addition to monitoring for the presence of heavy metals in whole fish, FWS monitors organic compounds, including several pesticides that have been banned.

The record is similar to toxic chemicals. FWS found that fish-tissue concentrations of DDT and related compounds, as well as dieldrin, declined by more than 60 percent between 1970 and 1986. The Great Lakes Program has amassed extensive data on pesticide concentrations in animal tissue. The data show that pesticides such as DDT have fallen more than 90 percent since the 1970s. Figure 10 depicts long-term trends in DDT concentrations in Lake Michigan trout (U.S. CEQ 1995, 491).

REGULATORY SYSTEM AND POLLUTION LEVELS

A fundamental question implicit in all the above discussion is to what extent changes in pollution levels are attributable to pollution control programs. It is neither conceptually nor factually correct to assume that, because declines in many pollutants have followed investment in pollution control programs, the decline is due to the programs. In a situation where multiple factors are at work, it cannot be assumed that one thing caused another because one followed the other in time.

Previous studies, notably those by Broder (1986) and MacAvoy (1987), have found little evidence of an association between pollution control investments and air quality. Indeed, the above data, particularly on air quality, suggest that other factors such as climate and industrial activity significantly affect environmental quality. For example, casual observation of Rust-Belt cities such as Pittsburgh suggests that air quality has improved dramatically over the past few decades; however, improvement is often attributed to the decline in heavy manufacturing rather than to laws and programs to control pollution.

In an effort to statistically disentangle the extent to which regulation may affect environmental quality, we analyzed economic, meteorological, and air pollution control cost data for a twenty-year period in three Rust Belt metropolitan counties: Allegheny, Pennsylvania (which includes the city of Pittsburgh), Baltimore, Maryland (Baltimore), and Cuyahoga, Ohio (Cleveland) (Powell 1997).

Our analysis was limited to data measured at ambient air monitoring sites. The counties were selected because they are metropolitan areas where the level and composition of economic activity substantially changed concurrently with the implementation of national air quality standards. In all three cities, steel manufacturing was, and to some extent remains, an important industry. If the effects of mandated pollution control investments can be detected in urban areas where economic changes have been stark, then we arguably would have more confidence that significant regulatory effects have occurred where changes in the level and composition of economic activity have been less dramatic.

Overall, the results of the analysis suggest that mandated pollution control investments have often had a significant effect in reducing air pollution levels. The effects of regulatory controls, however, generally have been overshadowed by the effects of economic changes, weather, and other factors. We also found that local factors, such as the level of manufacturing activity and local pollution control investments, fail to account for a majority of the variation in local air quality. The finding underscores the importance of regional or national factors (both regulatory and non-regulatory) in determining local air quality.

CONCLUSION

Data to illustrate long-term, national emissions trends, concentration in the environment, and human and environmental health effects are lacking for most environmental pollutants and media. Air pollution is the only category where systematic data exist. Based on the data, it is clear that pollution released to air from factories and individual automobiles has declined. In contrast with air data, information to illustrate the quality of the nation's water is in short supply. Construction of sewage treatment plants and the growth of programs to curb the use of contaminants such as phosphorous in commercial products have helped to improve water quality in some of the nation's rivers, lakes, and streams. The record on solid waste is less encouraging: economic prosperity and population growth have significantly expanded the amount of nonhazardous refuse we send to landfills and incinerators. Trends in hazardous waste are harder to interpret, primarily because the data are so poor. EPA is working with states and industry to improve the quality of hazardous waste information. Finally, a few stunning successes have occurred with toxics and pesticides, most notably in the release and application of heavily regulated or banned substances such as PCBs, mercury, and lead. While the substances are no longer emitted in large amounts, we lack data to determine how much of them remains in the environment.

REFERENCES

Adler, Robert W., Jessica C. Landman, and Diane M. Cameron. 1993. *The Clean Water Act: 20 Years Later*. Washington, D.C: Island Press.

Allen, D. T., and R. Jain, eds. 1992. *Hazardous Wastes and Hazardous Materials* 9(1): 1–111.

Broder, Ivy E. 1986. *Ambient Particulate Levels and Capital Expenditures: An Empirical Analysis*. Undated paper, Department of Economics, American University. As cited in Adler, Landman, and Cameron.

C&EN (*Chemical and Engineering News*). 1985. Production by the U.S. Chemical Industry, 1974–1984. June 10, p. 27.

———. 1990. Production by the U.S. Chemical Industry, 1969–1979. June 9, p. 38.

———. 1996. Production by the U.S. Chemical Industry, 1985–1995. June 24, p. 42.

Conservation Foundation and Research Institute for Public Policy. 1990. *Great Lakes, Great Legacy?* Baltimore: Conservation Foundation.

Constance, S., B. Sullivan, and M. E. Barron. 1989. Acute Illnesses among Los Angeles County Lifeguards According to Worksite Exposures. *American Journal of Public Health* 79(11). As cited in Adler, Landman, and Cameron.

Davies, J. Clarence. 1970. *The Politics of Pollution.* New York: Pegasus Press.

Dennis, Robin. 1995. Using the Regional Acid Deposition Model to Determine the Nitrogen Deposition Airshed of the Chesapeake Bay Watershed. In *Atmospheric Deposition to the Great Lakes and Coastal Waters.* Forthcoming. Washington, D.C.: Society of Environmental Toxicology and Chemistry (SETAC).

Dodell, David, ed. 1994. Blood Lead Levels: United States, 1988–1991. *MEDNEWS.* Health Info-Com Network (HICNet) Medical Newsletter. http://ch.nus.sg/MEDNEWS/aug94/7361_6.html.

Gianessi, Leonard P., and Henry M. Peskin. 1981. Analysis of National Water Pollution Control Policies 2. Agricultural Sediment Control. *Water Resources Research* 17(4): 803–21.

———. 1984. An Overview of RFF Environmental Data Inventory: Methods, Sources and Preliminary Results. Volume 1. Submitted to the Water Resources Division, USGS, in fulfillment of Purchase Order 000089.

Gottlieb, Robert, ed. 1995. *Reducing Toxics: A New Approach to Industrial Decisionmaking.* Washington, D.C.: Island Press.

ITFM (Intergovernmental Task Force on Monitoring Water Quality). 1994 The Strategy for Improving Water-Quality Monitoring in the United States. Reston, Virginia: USGS.

Jasinski, Stephen M. 1994. *The Materials Flow of Mercury in the United States.* Information Circular 9412. Washington, D.C.: U.S. Bureau of Mines, U.S. DOI.

Knopman, Debra S., and Richard A. Smith. 1993. Twenty Years of the Clean Water Act. *Environment* 35(1): 17–41.

Levine, W. C., W. T. Stephenson, and G. F. Craun. 1990. Waterborne Disease Outbreaks, 1986–1988. *Morbidity and Mortality Weekly Report* 39(SS-1).

Llewellyn, Thomas O. 1994. *Cadmium (Materials Flow).* Information Circular 9380. Washington, D.C.: Bureau of Mines, U.S. DOI.

MacAvoy, Paul. W. 1987. The Record of the Environmental Protection Agency in Controlling Industrial Air Pollution. In R. L. Gordon, H. D. Jacoby, and M. B. Zimmerman (eds.), *Energy Markets and Regulation.* Cambridge, Massachusetts: Ballinger.

MDEQ (Michigan Department of Environmental Quality). 1996. *State of the Great Lakes: Annual Report for 1995.* Office of the Great Lakes. Lansing: MDEQ.

NAPA (National Academy of Public Administration). 1995. *Setting Priorities, Getting Results: A New Direction for EPA.* Washington, D.C.: NAPA.

NRC (National Research Council). 1993. *Measuring Lead Exposure in Infants, Children, and Other Sensitive Populations.* Washington, D.C.: National Academy Press.

NRDC (Natural Resources Defense Council). 1996. *Testing the Waters.* Vol. 6. New York: NRDC.

NSTC (National Science and Technology Council). 1997. National Environmental Monitoring and Research Workshop Proceedings. (Workshop held September 25–27, 1996). February 25. Washington, D.C.: NSTC.

O'Connor, T. P. 1991. Concentrations of Organic Contaminants in Mollusks and Sediments at NOAA National Status and Trend Sites in the Coastal and Estuarine U.S. *Environmental Health Perspectives* 90: 70, 73.

Papp, John. 1994. *Chromium Life Cycle Study.* Draft Information Circular. Washington, D.C.: U.S. Bureau of Mines, U.S. DOI.

Powell, Mark. 1997. *Three-City Air Study.* Discussion Paper 97-29. Washington, D.C.: Resources for the Future.

Probst, Katherine N., Don Fullerton, Robert E. Litan, and Paul R. Portney. 1995. *Footing the Bill for Superfund: Who Pays and How?* Washington, D.C.: Brookings Institution.

Roque, Julie. 1995. Personal communication with the author, November. (Roque was a senior policy analyst in the White House Office of Science and Technology Policy.)

Russell, Clifford. 1990. Monitoring and Enforcement. In Paul R. Portney (ed.), *Public Policies for Environmental Protection.* Washington, D.C.: Resources for the Future.

Schmitt, C. J., and W. G. Brumbaugh. 1990. National Contaminant Biomonitoring Program: Concentrations of Arsenic, Cadmium, Copper, Lead, Mercury, Selenium, and Zinc in U.S. Freshwater Fish, 1976–1984. *Archives of Environmental Contamination and Toxicology* 19: 731–47.

Smith, Richard A., Richard B. Alexander, and Kenneth J. Lanfear. 1991. Stream Water Quality in the Conterminous United States: Status and Trends of Selected Indicators during the 1980s. USGS Water Supply Paper 2400. Washington D.C.: U.S. GPO.

Smith, Richard A., Richard B. Alexander, and Gordon Wollman. 1987. *Analysis and Interpretation of Water Quality Trends in Major U.S. Rivers, 1974–1981.* USGS Water Supply Paper 2307. Washington D.C.: U.S. GPO.

U.S. CEQ (U.S. Council on Environmental Quality). 1990. *Environmental Quality: The 21st Annual Report.* Washington, D.C.: U.S. GPO.

———. 1995. *Environmental Quality: The 24th Annual Report.* Washington, D.C.: U.S. GPO.

———. 1996. *Environmental Quality: The 25th Anniversary Report.* Washington, D.C.: U.S. GPO.

USDA (Department of Agriculture). 1994. *Agricultural Resources and Environmental Indicators.* Agricultural Handbook No. 705. Natural Resources and Environment Division, Economic Research Service. Washington, D.C.: USDA.

U.S. DOC (Department of Commerce). 1991. *The 1990 National Shellfish Register of Classified Estuarine Waters.* National Oceanic and Atmospheric Administration, National Ocean Service, Office of Oceanography and Marine Assessment, Strategic Assessment Branch. Washington, D.C.: U.S. DOC.

U.S. DOE (Department of Energy). 1991. *Industrial Waste Reduction Program: Program Plan.* Washington, D.C.: Office of Industrial Technology.

U.S. DOI (Department of the Interior). 1992. Data on chromium and nickel production. In *Minerals Yearbook.* Washington D.C.: Bureau of Mines, U.S. DOI.

U.S. EPA (Environmental Protection Agency). 1984. *Revised Evaluation of Health
Effects Associated with Carbon Monoxide: An Addendum to the 1979 EPA Air Qual-
ity Criteria Document for Carbon Monoxide.* Office of Health and Environmental
Assessment. Washington, D.C.: U.S. EPA.

————. 1986. *Second Addendum to Air Quality Criteria for Particulate Matter and Sulfur Oxides (1982): Assessment of Newly Available Health Effects Information.* Office
of Health and Environmental Assessment. Washington, D.C.: U.S. EPA.

————. 1988. *Summary of Selected New Information on Effects of Ozone on Health and
Vegetation: Draft Supplement to Air Quality Criteria for Ozone and Other Photo-
chemical Oxidants.* Office of Health and Environmental Assessment. Washington, D.C.: U.S. EPA.

————. 1990. *Environmental Investments: The Cost of a Clean Environment.* Report of
the Administrator of the Environmental Protection Agency to the Congress of
the United States. Office of Policy, Planning, and Evaluation. Washington,
D.C.: U.S. EPA.

————. 1991. *EPA's Pesticide Programs.* 21T-1005. Office of Pesticides and Toxic Substances. Washington, D.C.: U.S. EPA.

————. 1994a. *A Clean Air Act Exposure and Effects Assessment 1993–1994: A Prototype Biennial Ecological Assessment.* Office of Research and Development.
Washington, D.C.: U.S. EPA.

————. 1994b. *National Water Quality Inventory: 1992 Report to Congress.* Office of
Water. Washington, D.C.: U.S. EPA.

————. 1994c. *Technical Analysis of Response of Chesapeake Bay Water Quality Model to
Loading Scenarios.* Chesapeake Bay Modeling Subcommittee. Washington,
D.C.: U.S. EPA.

————. 1994d. *Trends in Phosphorus, Nitrogen, Secchi Depth and Dissolved Oxygen in
Chesapeake Bay, 1984 to 1992.* Chesapeake Bay Program. Washington, D.C.:
U.S. EPA.

————. 1995a. *A Conceptual Framework to Support the Development and Use of Environmental Information for Decision-Making,* EPA 230-R-95-012. Office of Policy,
Planning, and Evaluation. Washington, D.C.: U.S. EPA.

————. 1995b. *National Public Water System Supervision Program FY1993 Compliance
Report.* Office of Water. Washington, D.C.: U.S. EPA.

————. 1995c. *National Air Pollutant Emission Trends, 1900–1994.* Office of Air Quality Planning and Standards Research. Research Triangle Park, North Carolina: U.S. EPA.

————. 1995d. *National Water Quality Inventory: 1994 Report to Congress.* Office of
Water. Washington, D.C.: U.S. EPA.

————. 1995e. *Toxics Release Inventory: Public Data Release.* Office of Pollution Prevention and Toxics. Washington, D.C.: U.S. EPA.

———. 1996a. Science Advisory Board. Letter from Dr. George T. Wolff, Chair, Clean Air Scientific Advisory Committee, to Carol M. Browner regarding "Closure by the Clean Air Scientific Advisory Committee (CASAC) on the Draft Air Quality Criteria for Particulate Matter." March 15. On file with authors.

———. 1996b. *Characterization of Municipal Solid Waste in the United States: 1995 Update*, EPA530-R-96-001. Solid Waste and Emergency Response. Washington, D.C.: U.S. EPA.

———. 1996c. *National Air Pollutant Emissions Trends, 1900–1995*. Office of Air Quality Planning and Standards Research. Research Triangle Park, North Carolina: U.S. EPA.

———. 1996d. *National Air Quality and Emissions Trends Report, 1995*. Office of Air Quality Planning and Standards Research. Research Triangle Park, North Carolina: U.S. EPA.

———. 1996e. *National Biennial RCRA Hazardous Waste Reports, 1989–1993*. Office of Solid Waste and Emergency Response. Executive summary files for reporting years 1991 through 1993. http://www.epa.gov/OSWRCRA/hazwaste/data.

———. 1996f. *Public Water System Inventory and Statistics: FY1991–FY1995*. Office of Water. Washington, D.C.: U.S. EPA.

———. 1996g. *1994 Toxics Release Inventory: Public Data Release*. Office of Pollution Prevention and Toxics. Washington, D.C.: U.S. EPA.

———. 1997a. *EPA Releases Plan for the Waste Information Initiative*. Office of Public Information. Friday, February 21. Washington, D.C.: U.S. EPA.

———. 1997b. *1995 Toxics Release Inventory: Public Data Release*, Document 745-R-97-005. Office of Pollution Prevention and Toxics. Washington, D.C.: U.S. EPA.

U.S. FDA (Food and Drug Administration). 1996. *Pesticide Residue Monitoring Program: Eighth Annual Report*. http://vm.cfsan.fda.gov/~frf/94prmr.html.

U.S. GAO (General Accounting Office). 1991. *Waste Minimization Data Are Severely Flawed*, GAO/PEMD-91-21. Washington D.C.: U.S. GAO.

U.S. ITC (International Trade Commission). 1970–1993. *Synthetic Organic Chemicals: United States Production and Sales* and reports for individual years 1991, 1992, 1993. Washington, D.C.: U.S ITC.

Wholey, Joseph, J. W. Scanlon, H. G. Duffy, J. S. Fukomoto, and L. M. Vogt. 1970. *Federal Evaluation Policy: Analyzing the Effects of Public Programs*. Washington, D.C.: Urban Institute.

Woodbury, William D., Daniel Edelstein, and Stephen M. Jasinski. 1993. *Lead Materials Flow in the United States 1940–1988*. Draft. Washington, D.C.: Bureau of Mines, U.S. DOI.

6

Targeting the Most
Important Problems

If the overarching goal of the pollution control system is to reduce risk, one criterion for measuring its effectiveness is to examine whether the resources devoted to controlling pollution are targeted at the most pressing environmental problems. Comparative risk assessment is a tool to help decisionmakers better understand whether the system is trained on the right goals. This chapter examines how the method has been applied to evaluate EPA's priorities, discusses areas where the tool must be improved, and then applies some of the findings from prominent comparative risk assessments to examine whether the goals of the pollution control system are trained on the problems that pose relatively high human and environmental health risks.

The pollution control system has reduced some of the most visible threats to human health, such as smoke from factories and sewage in surface water. However, when contrasted with cost and spending data that illustrate how EPA and society allocate resources to control pollution, comparative risk exercises suggest that more elusive but high risks to humans and ecosystems—such as those resulting from indoor radon and water pollution that runs off city streets and farm fields—remain to be addressed. Before examining whether the EPA is directing its resources toward high-risk problem areas, we explain why decisionmakers need tools to maximize the usefulness of limited resources.

THE NEED TO SET PRIORITIES

Since 1970, the United States has spent more than $1 trillion in the effort to control pollution (Jaffe and others 1995). While the U.S. citizenry consistently shows support for spending dollars to achieve environmental progress, the growing perception is that policymakers must increasingly target resources wisely. Historically, priority setting has not been a part of the pollution control system. The Clean Air Act has been interpreted to prevent the EPA administrator from weighing the environmental benefits of risk reduction against the costs of pollution control when setting national ambient standards. Other statutes, such as the Toxic Substances Control Act (TSCA), explicitly require EPA to balance the potential benefits of regulatory risk reduction efforts against costs. More recently, the idea that decisionmakers must carefully consider the risk reduction and economic impacts of their policies has gained momentum due to several factors.

Perhaps the most important of these factors has been the reality of budget-tightening occurring at all levels of government. The shrinking of revenue is more a political than an economic phenomenon. Priority setting also has been encouraged by awareness that for U.S. companies to be competitive in the world economy, regulatory efforts should not be wasted on unimportant problems.

Private firms and states and localities increasingly maintain that federal mandates often impose excessively costly pollution control requirements. Localities in particular have turned to comparative risk assessment as a way to identify important problems and develop strategies that are more effective and efficient than "one-size-fits-all" federal pollution control laws.

While some of the factors that contributed to the push for priority setting are of recent origin, the analytic concept of resource finitude has been around for some time. "Resources" are simply things such as minerals, timber and food, and secondary materials that we use to sustain us (Landsberg, Fischman, and Fisher 1963, 3). Depending on the adjective, the term also can refer to things that a human, a factory, or even a government agency such as EPA needs to use in order to function. Resources are not simply things that are traded and carry a market value. They also can include very fundamental, but natural, elements such as clean air and clean water.

The concept of resource scarcity is central to public policy because it suggests that decisionmakers must allocate resources to reduce risks in a way that is cost-effective. If society had infinite amounts of the land, labor, and machinery needed to make and consume other things, we could reduce pollution at any cost. If natural resources such as clean air or clean water were infinite, pollution would not be a problem. Typically, however, society must divert some resources previously used for making things or addressing other problems toward the task of controlling pollution. The critical task of such decisions is understanding how to apply resources to best reduce risks to humans and the natural environment.

PRIORITIES

In simple terms, a list of priorities is just a set of things we would prefer to have or to do. Priorities can be ranked numerically or in relative terms like "high," "medium," and "low." Since resources are finite, most of us cannot have or do everything. Our preferences are constrained by income or by our budget. How people allocate money is one indicator of their priorities. Of course, not everything we think of as important is immediately revealed through spending patterns. For example, "spending time with family" may not show up in the family budget each month, but it is possible to convert the time spent into a money measure. While the practice of assigning monetary values to some things may be controversial, constraints—whether in the form of time, money, or energy—are an inescapable fact of life.

The same rules apply to government agencies. Agencies have many problems to solve and limited resources with which to solve them. Should the manager spend money on high-profile, low-risk problems, or on low-risk, high-profile problems? There is no easy answer. Ideally, trying to determine where the resources should go would incorporate information about public risk perceptions, resource constraints, and estimates about probable adverse human and environmental health effects.

Due to constraints on time and resources, we are unable to provide original data on public risk perceptions here. Some of the comparative risk exercises referenced later in this chapter begin to do just that. We do attempt to develop a broader picture of priorities from expenditure data and risk assessment information.

In the past, estimates of EPA priorities have focused on government expenditures only. Such an analysis reveals what Congress, or EPA, in response to public pressure, may view as being important environmental problems. But such approaches fail to distinguish how environmental laws affect different parts of society. For example, an industry diverts investments from other opportunities, like expanding a plant or upgrading a lunchroom, when it invests in technological equipment or labor time spent on regulatory paperwork. Households, in turn, pay for laws in the form of taxes to run federal agencies as well as in higher product prices, when firms pass the cost of pollution control to consumers. Some federal laws also require states and localities to invest heavily in programs to improve environmental quality, such as construction of sewage treatment plants and measures to provide safe drinking water. For each of these groups, controlling pollution carries a cost in the form of diverted resources or lost opportunities.

A first step in evaluating whether goals are trained on the right problems is to examine how society and EPA currently allocate resources to control pollution. The first part of this discussion focuses on the historic and current costs of pollution control to society (the term "society" refers to the private sector, including firms and households, and all levels of government, including EPA). When we contrast societal expenditures with EPA expenditures, a minor overlap is evident because a small portion (see Figure 1) of the societal expenditures are EPA expenditures.

The primary source of annual data on environmental spending patterns to EPA, other federal entities, and the private sector is generally referred to as the *Cost of Clean* (U.S. EPA 1990a). While it is the central source of data on environmental costs, the data are highly unreliable because of the way in which they are collected, computed, and reported. The total costs of pollution control in the United States can appear to vary significantly, depending on the source of the data and the assumptions built into the analysis.

To illustrate the potential magnitudes of difference among the cost tabulations, consider that, according to EPA, the projected total cost of pollution control to society in 1997 is nearly $200 billion (U.S. EPA 1990a). Other estimates, using different assumptions and computational methods, place the total cost at around $100 billion per year (NAPA 1995). The actual magnitude is likely to fall somewhere in-between. The reader is

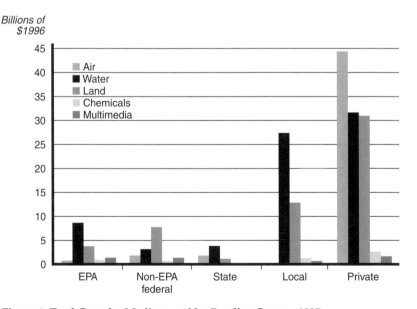

Figure 1. Total Costs by Medium and by Funding Source, 1997.

Source: U.S. EPA 1990a, Table 8-12A, 8-51.

Note: Present implementation cost of regulations, annualized at 7%.

therefore cautioned to interpret the data only as rough guides to how agencies and sectors allocate resources to environmental problems.

One reason *Cost of Clean* may overstate actual costs is that the data reported since 1986 are based not on actual spending but on forecasts constructed from historic spending patterns. For each year following 1986, forecasts are reported as annualized costs. Annualized costs do not reflect costs in the year they occur, but add the sum of operating costs for the year in question plus amortized capital costs, which include interest and depreciation associated with accumulated capital investment (U.S. OTA 1995). Since the assumed life of capital equipment varies, *Cost of Clean* uses different amortization schedules that reflect various assumptions on the life of the capital investment. They are reported at depreciation rates of 3, 7, and 10 percent of annualized expenditures. Most references to *Cost of Clean* use data based on the 7 percent category; however, some analysts think the 7 percent estimates overstate the true costs of pollution control (U.S. OTA 1995).

While some of the cost data were developed at EPA, most were originally collected from surveys collected by the Census Bureau and reported by the Department of Commerce's Bureau of Economic Analysis (BEA). In contrast to EPA, the latter counts capital costs and expenditures in the years that they are made. The surveys ask firms and other government agencies to estimate operations and maintenance costs and capital expenditures on pollution control, and thus omit other important costs that are not as apparent or are difficult to measure. Examples include the costs of new equipment designed not specifically for pollution control that nonetheless yields superior environmental performance, or of managerial time spent reporting on environmental compliance. In contrast with EPA, BEA estimates only report a small fraction of the cost of solid waste management and collection and a smaller share of Superfund costs.

Finally, some semantic clarifications regarding the term "cost" are in order. Economists distinguish between direct expenditures necessitated by regulation and the true social costs of regulation. The data here primarily deal with expenditures and not social costs. Efforts to evaluate the true social costs associated with existing regulations begin with expenditure data, then trace the effects of the regulation throughout the economy by employing sophisticated models (Hazilla and Kopp 1986). For example, Hazilla and Kopp's estimates of true social costs between 1975 and 1985 are initially much lower than EPA expenditure data ($6.8 versus $14.1 billion in 1975). By 1985, EPA data and estimates of true social costs are $24.6 and $33 billion, respectively, a reversal of the 1975 situation. Hazilla and Kopp attribute the differences in part to the possibility that consumer prices tended to rise as regulations became more widespread and more firms passed the costs on to buyers. When regulations targeted relatively fewer firms, consumers simply substituted less expensive products manufactured by unregulated firms for products made by regulated entities. In addition to the data reported in *Cost of Clean*, we also use an additional source, the summary of EPA's 1998 budget (U.S. EPA 1997). The primary shortcoming of the budget estimates is that, for the most part, they fail to itemize spending on individual problems such as indoor air pollution or radon that are part of broader program categories. In the next few years, it will no longer be possible even to evaluate spending by media category. EPA spending data by agency program has come from the Office of Management and Budget; however, Congress restructured the agency's accounts in 1996 in order to move away from a process that targeted funds toward individual air, water, and land-based programs (U.S. OMB 1996, 873). As a result, OMB data will no longer report EPA expenditures by individual media.

EPA cost projections appear in Table 1. The rows roughly correspond to EPA media programs. In order to help compare cost data with priorities identified by comparative risk exercises, we have created three arbitrary spending categories titled "high," "medium," and "low" to designate in relative terms how much society spends on each environmental problem area.

In addition to differences in the ways in which the data were computed, there are slight variations in how they are reported. Data in the *Cost of Clean* are reported in categories that roughly correspond to EPA programs: air and radiation, water, land, and chemicals. In cases where there is overlap or the activity is considered to be "pollution prevention," the category is reported as "multimedia."

Table 1. Percentage of Pollution Control Cost to Society by Problem Area, 1997.

Problem rank	Projected cost
High	Water quality (37%)
	Air (28%)
Medium	Solid waste (15%)
Low	Hazardous waste (7%)
	Drinking water (4%)
	Superfund (4%)
	Leaking underground storage tanks (2%)
	Pesticides (1%)
	Toxic substances (0.8%)
	Radiation (0.5%)

Note: EPA data are based on projected costs at 7% annualized rate assuming full implementation in 1997 reported in $1996.

Sources: U.S. EPA 1990a, Tables 3-3, 4-3, 5-3, 6-3, 8-6; U.S. OTA 1995, 49.

Table 1 illustrates that surface water quality comprises the largest portion of pollution control costs to society (about 37 percent). Expenditures to improve water quality outrank other categories because they include the cost of "water infrastructure" programs devoted to the construction and improvement of publicly owned sewage treatment systems.

After water, air quality protection comprises the next largest share of costs. *Cost of Clean* shows that solid waste comprises about 15 percent of the total cost of pollution control to society, while hazardous waste comprises about 7 percent. Problems that comprise 5 percent or less of the total cost include drinking water, Superfund, and pesticides. In all cases, spending devoted to controlling toxic substances or radiation comprise 1 or less than 1 percent of the total cost of pollution control to society.

To better illustrate how sectors allocate spending among different environmental problems, Figure 1 presents these data by sector and by environmental medium. Here, the reporting categories from *Cost of Clean* have been condensed somewhat from those presented above. The category "land" has been designated to include Superfund, hazardous waste, pesticides, and solid waste management. "Chemicals" refers to investments under the Toxic Substances Control Act. As mentioned, "multimedia" refers to categories of overlap and initiatives devoted to pollution prevention. As Figure 1 illustrates, the private sector overwhelmingly pays the largest share of total costs, especially for compliance with the air, water and land categories (U.S. EPA 1990a, 8-51). Localities pay for a similar share of surface water quality improvement costs, and a smaller share of costs that fall into the land category. States spend the largest portion of their total funds administering programs to improve surface water quality (U.S. EPA 1990a, 8-51). In the non-EPA federal category (primarily the Departments of Defense and Energy), funding is devoted to cleaning up and managing contaminated military and nuclear weapons sites. The largest portion of EPA's projected expenditures in 1997 is devoted to improving surface water quality (U.S. EPA 1990a, 8-51).

TOTAL COST OF POLLUTION CONTROL TO EPA

EPA's estimated $6.8 billion budget for FY1997 is a small portion of the *projected* $198 billion cost of pollution control to society in 1997 (U.S. EPA 1997, 80). Comparisons are presented in tabular form on the following pages.

Table 2 compares projections with actual EPA budget outlays for selected media in 1997. The table presents the costs on a percentage basis. It shows that EPA devotes relatively similar proportions of the budget to the same categories of problems as society at large. Both society and EPA spend the most on improving surface water quality, followed by air and

Table 2. Percentage of Total Cost of Pollution Control to Society and to EPA in 1997, by Program.

Media	1997 total projected cost to society[a]	FY 1997 estimated EPA expenditures
Water quality[b]	37%	53.79%
Superfund	4%	28.91%
Air[c]	28%	5.43%
Hazardous waste[d]	22%	3.73%
Drinking water	4%	2.04%
Pesticides	1%	2.25%
Toxic substances	0.8%	1.69%[e]
Leaking underground storage tanks	2%	1.27%
Radiation	0.5%	0.39%
Indoor air pollution[f]	n/a	0.15%
Radon[f]	n/a	0.36%

Note: Excludes the following expenditures: State, Local, and Tribal Grants; Buildings and Facilities; Science and Technology; Office of Inspector General; Multimedia; and Oil Spills. When these are factored out, the current estimate for FY1997 budget comes to $4.663 billion.

[a]1997 projections represent costs incurred to achieve full implementation of existing regulations annualized at 7%. Data are reported in $1996.

[b]Cost of Clean estimates for water quality include state revolving funds for construction grants. In order to compare the data, these amounts were also taken from the EPA Summary and included in actual EPA 1997 outlays.

[c]Air and radiation cost for the 1997 projections of the EPA component of total society costs were not separated in Cost of Clean. It was assumed that they made up the percentage of costs as the actual 1997 outlays.

[d]EPA budget estimates aggregate solid and hazardous wastes into a single hazardous waste category. Thus, Cost of Clean categories for solid 5.1 and hazardous 5.2 wastes also are added here to aid in comparison.

[e]Estimate for Toxics Program ($0.078 billion) excludes management and support ($0.481 billion).

[f]Data obtained through personal communication, Office of Radiation and Indoor Air, November, 1996. The percentages have remained roughly unchanged in recent years.

Sources: U.S. EPA 1990a, Tables 3-3, 4-3, 5-3, 6-3, 7-3, 8-6; U.S. EPA 1997, 80–81.

waste (hazardous, solid, and Superfund). When taken alone, EPA's estimate for infrastructure financing in 1997 was about $2.2 billion, while funding for other programs to enhance water quality such as addressing nonpoint source problems came to approximately $272 million (U.S. EPA 1997, 80).

Radiation problems rank among the lowest in funding from both society as a whole and EPA. Part of the reason that EPA's funding levels for radiation problems are low is that many aspects of radiation are handled by other federal agencies, including the Department of Energy and the Nuclear Regulatory Commission.

Overall, the chief difference between societal costs and EPA's costs is one of magnitude and not program priority. Surface water quality ranks highest for both society and for EPA at 37 and 54 percent, respectively. Superfund is one notable exception: EPA's 29 percent estimate of spending on Superfund greatly exceeds the projected social cost of 4 percent reported in *Cost of Clean.* Total cost to society for Superfund may be low because cleanup only involves a relatively small number of sites. Government pays a high percentage of the cost because it absorbs the cost when EPA is unable to place liability on a company. Part of the difference also may be political: Superfund sites pose relatively low risks to the public in general, yet poll after poll reveals the public's concern with hazardous waste sites. The difference may therefore also reflect EPA's response to public pressure, rather than actual human health risk. Air quality is another area where percentages vary. EPA devotes a smaller percentage to air problems than society does because the Clean Air Act places the primary obligation for compliance expenditures on the private sector, as shown in Figure 1.

Two items failing to appear in the FY1997 budget estimates are indoor radon and indoor air pollution. Indoor radon is not regulated by EPA at all and only minimally by the states. The EPA radon program is 0.36 percent of the agency's budget (U.S. EPA 1996). Similarly, indoor air pollution, the second-ranked health problem, is for the most part neither regulated by EPA nor by the states (U.S. EPA 1996). Local building codes address some aspects of the indoor air problem, such as rates of air exchange. EPA has an indoor air program—it is 0.15 percent of the agency's budget (U.S. EPA 1996).

If expenditures on environmental problems are taken as a proxy for priorities, then problems such as surface water pollution appear to rank highest, followed by air pollution and hazardous waste. Problems such as toxic chemicals, pesticides, and radiation receive relatively small amounts of funding across society and in EPA's budget. These trends are reflected in both projections of total societal cost and EPA budget data. Examined from the perspective of who pays for these costs, data show that the private sector picks up the largest share of these costs, followed by local government, states, other federal government agencies, and finally EPA. The only striking difference between EPA's budget data and societal projections arises in the area of Superfund, where the percentage of EPA's actual budget outlays exceeds the cost to the private sector as projected in *Cost of Clean.* In the following section we will examine several methods with which to compare current spending priorities with relative risks posed by environmental problems.

METHODS FOR EVALUATING EPA'S PRIORITIES

When its budgetary outlays are compared with results from some prominent public opinion polls, EPA's priorities seem largely on target. Polls reveal that the public worries about hazardous waste facilities, abandoned hazardous waste sites, and chemicals housed in underground storage tanks (Main 1991). Some of these programs, such as Superfund, are among EPA's most expensive programs. Viewed from the perspective of public attitudes, Congress and EPA appear to be training resources on the proper priorities. However, when viewed in terms of total risk reduction and total benefit to the population, results change significantly.

In the following discussion, we briefly outline the methods employed for comparative risk assessment (CRA). We then illustrate the results of a few more prominent CRA exercises in order to reveal areas where EPA appears to be devoting significant resources to problems that pose relatively moderate to low risks to human health and the environment.

The term "comparative risk assessment" carries different meanings, but can refer to two different analytic exercises: "relative risk ranking" and "programmatic risk assessment" (Davies 1996, 5–6). Relative risk ranking refers to a quantitative analytic technique used to compare the likelihood of hazard between two or more relatively known substances or activities. In this book, we use the expression "CRA" (comparative risk assessment) in a broader sense to refer to a more qualitative method used to identify whether scarce, public resources are trained on the right pollution control problems. Also known as "programmatic risk assessment," this method involves both identifying and comparing a large number of risks.

To illustrate the strengths and inherent limitations of CRA, it is necessary to first define some terms and identify analytic predecessors to the methodology. The most central of these is the concept of "risk." In the context of environmental problems, "risk" is defined as the "likelihood that injury or damage is or can be caused by a substance, technology, or activity" (Davies 1996, 5). Risk is typically stated in terms of probabilities, for example, the probability that chemical X will cause a one-in-one-million excess lifetime risk of cancer. Many observers stress that in addition to a numeric output, "risk" also is a function of human perceptions, values, and even relative station in life. Risk perceptions may be colored by variables such as educational level or the extent to which individuals believe that they can control their exposure to a risk (Finkel and Golding 1994).

Risk assessment, from which CRA evolved, is a method to produce a numerical estimate of the magnitude of assessed risk. In the United States, risk assessment is generally understood as a four-step exercise, organized into the following tasks (NRC 1983):

1. Identify the type of injury that a substance or activity may cause.
2. Estimate the relationship between exposure to the substance or event and harm. Step 2 is typically administered on animals and the results extrapolated to humans through a number of calculations.
3. Estimate the potential population exposure to the hazard. Again, methodological difficulties require the analyst to make a number of simplifying assumptions on how target populations are exposed.
4. Combine Steps 2 and 3 to generate a characterization of the amount of injury that may be generated by the substance or activity.

The results of risk assessments provide important guides on how much potential hazard a substance or activity may pose. However, the process requires analysts to build in a number of assumptions because the underlying scientific data regarding adverse effects and exposure are limited.

For example, under Step 2, analysts must make a number of assumptions in order to extrapolate from observations on mice or rats that are fed high doses of a substance to human populations that may be exposed to relatively low doses. Such assumptions often produce estimates varying by several orders of magnitude. In the case of suspected carcinogens such as saccharine, one study posited that exposure would result in five excess cancer cases per million exposed, while another found the substance caused 1,200 excess cancer cases per million exposed. At the other extreme, an industry-sponsored study found that exposure to the substance only led to one excess cancer case per *billion*. Clearly then, the results of risk assessments do not represent the "true" or actual likelihood of an event, but rather reflect estimates of the probability that an adverse effect will result.

Programmatic CRAs are informed by data generated from risk assessments. The advantage of the tool is that it provides decisionmakers with a guideline to understand how various problems may affect human health and the environment. Because the underlying scientific data generates highly uncertain risk estimates, the CRA process requires constant updating and refinement as the scientific data evolve. While a CRA can provide guidelines on how to target resources more effectively, it cannot tell decisionmakers how to make difficult trade-offs such as whether to direct scarce resources toward reducing childhood cancers or diseases among the elderly. CRAs should be understood as a tool that contributes information to a decision rather than a rule by which decisions must be made. Other essential factors in decisions include cost, equity, and implementability.

Just as there is no single definition of a CRA, nor one exact estimate of the degree of risk carried by a substance or activity, there is no single "best" method for conducting a programmatic risk assessment. In some instances, the exercises are structured to incorporate more impressionistic aspects of risk such as dread and familiarity by pairing public participants with technical experts. Such processes are as much art form as science. They involve analysis, discussion, and often debate. The exercises can range in duration from several months to well over a year. When complete, the output is typically a broad ranking of environmental priorities. While there is no widely agreed upon method to conduct programmatic CRAs, experts agree that the exercises typically involve problem definition and analysis, risk ranking, and priority setting, based on rankings (Minard 1995, 24).

While CRAs appear to be an excellent tool for identifying problems and defining strategies, most exercises to date have not yet resulted in significant changes in the way lawmakers and administrators actually target resources at environmental problems (Minard 1995, 24).

EPA and its Science Advisory Board have completed two large-scale programmatic CRAs. A third followup is currently under way. *Unfinished Business* was prepared in 1987 by a team of senior EPA analysts who identified thirty-one environmental problem areas and qualitatively ranked problems according to four categories: cancer, noncancer effects, ecological health, and social welfare. The effort revealed as much about agency priorities as the methodological weaknesses of the nascent tool (U.S. EPA 1987). In terms of priorities, the CRA revealed that problems EPA analysts identified as high and medium risks failed to correspond to public opinion and EPA's budget. According to *Unfinished Business,* high cancer risks to humans include:

- Worker exposure to chemicals
- Indoor radon
- Pesticide residues on food
- Indoor air pollution other than radon
- Consumer exposure to chemicals
- Hazardous/toxic air pollutants

High noncancer health risks include:

- Criteria air pollutants
- Hazardous air pollutants
- Indoor air pollutants (other than radon)
- Drinking water
- Accidental releases (toxics)
- Pesticide residues on food
- Application of pesticides
- Consumer product exposure
- Worker exposure to chemicals

Risks to ecosystems were reported by category. Category One risks include stratospheric ozone depletion and global warming. Category Two risks include physical alteration of aquatic habitats and activities that extract natural resources, such as mining and gas and oil removal. Under the final category, welfare effects, problems ranked by order of importance include:

- Criteria air pollution (including acid precipitation)
- Nonpoint source discharges to surface waters
- Indirect point source discharges to surface waters
- Discharges to estuaries, coastal waters, and oceans from all sources

Problems such as direct point source discharges to surface water from industrial and other sources ranked lowest as a social welfare problem.

Not surprisingly, the biggest challenge to EPA analysts in conducting the CRA involved the lack of scientific data to support their conclusions. In the absence of data, analysts relied on expert opinion. Another methodological difficulty was the somewhat arbitrary effect of isolating environmental problems by category. Such categorizations tend to over-simplify the complex, interrelated causes and manifestations of many environmental problems.

Related to the categorization problem is the question of how to assign numeric rankings to risks. In *Unfinished Business,* risks under the cancer and social welfare categories are assigned numeric values. However, analysts did not rank risks reported in the noncancer category because the risk data are limited, and comparing the various kinds of noncancer risks is inherently value-laden. Deciding whether to target scarce resources toward reducing childhood asthma or immune-deficiency risks requires both value judgments and more information than a single numeric output from a risk assessment.

Due in part to the methodological difficulties revealed by *Unfinished Business,* EPA's Science Advisory Board (SAB) in 1990 conducted its own attempt at ranking. *Reducing Risk* produced some results similar to those in *Unfinished Business,* but also showed the weakness of some of the data to support the earlier rankings. According to the SAB, human health risks that were firmly supported include:

- Ambient air pollutants
- Worker exposure to chemicals in industry and agriculture
- Indoor pollution
- Drinking water pollution

Risks to ecosystems supportable by underlying data include:

- Habitat alteration and destruction
- Species extinction and overall loss of biological diversity
- Stratospheric ozone depletion
- Global climate change

The SAB also ranked as medium risks the problems of water pollution and pesticides—problems that were ranked higher by *Unfinished Business*. The SAB did not rank the risks numerically.

In addition to refraining from numeric rankings, *Reducing Risk* reviewers omitted the social welfare category in part because they found that existing economic valuation tools tended to understate the value of natural resources. In response, the SAB recommended that EPA "develop improved methods to value natural resources and to account for long-term environmental effects in its economic analysis" (U.S. EPA 1990b).

Despite the limitations of CRA revealed by both *Unfinished Business* and the SAB's subsequent peer review of the study, the advisory board nonetheless concluded that CRA is an important tool for identifying priorities and that its methods should continue to be refined (U.S. EPA 1990b). One way EPA has sought to improve the power of the tool has been to encourage regional, state, and local efforts to capture geographic variation lost when analysts must generalize about how risks might affect the entire nation. In concrete terms, ecological risks such as discharges to coastal waters are not much of a problem in interior states; air pollution problems associated with high concentrations of automobiles or industrial sources are not much of a problem in remote, rural areas.

To date, each of the ten EPA regions has completed such exercises and EPA has sponsored similar efforts in some twenty-five states and thirteen municipalities and tribal territories. Figures 2 through 5 depict the results of regional and state efforts. Regional CRAs have produced results that are not numerically ranked, but merely reported as "high," "medium," and "low." Some states have combined the various categories into an integrated risk ranking.

As these figures indicate, some rankings are quite similar to the results reported by their two federal CRA predecessors. In terms of risks to human health, all regions identified indoor radon as a high risk. Indoor air problems other than radon were identified as a high risk by eight regions. These problems were followed by pesticides and hazardous air pollutants. Problems identified as high risk by only one region include accidental chemical releases, and problems associated with carbon dioxide (CO_2) and global warming.

All regions identified physical deterioration of ecosystems on land as a high-risk problem. Physical degradation of water and wetland habitats followed. Only one region identified each of the following as posing high

Percent of
EPA regions
ranking risk
as high

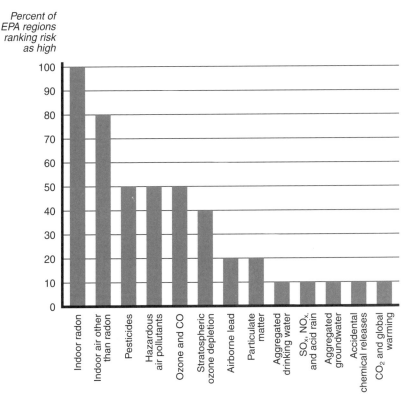

Figure 2. Human Health Risks Ranked as High by EPA Regions.
Source: U.S. EPA 1993, 20.

risks to ecosystems: particulate matter, hazardous air pollutants, and groundwater.

Reporting rankings for six of the states and localities where EPA sponsored CRAs is complicated by the fact that each conducted and consequently reported its programmatic assessments differently. Not all states reported on the same problems. Figures 4 and 5 present in graphic form the number of states and localities that ranked the problem.

Some similarities exist between the rankings reported by states and localities and those reported by EPA analysts in *Unfinished Business.* For example, indoor air pollution and indoor radon problems rank as the highest risks to human health. The biggest departure from *Unfinished Business* occurs in the human health category. The jurisdictions rated problems associated with pesticides and chemical exposure lower than did *Unfinished Business* and its successor, *Reducing Risk.*

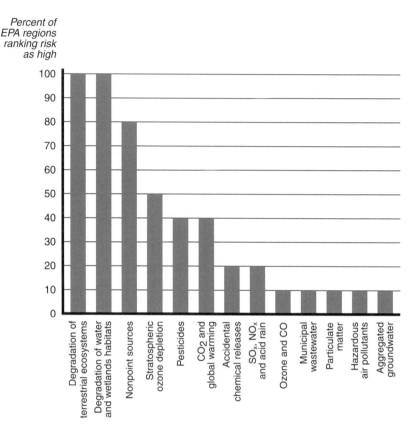

Figure 3. Ecological Health Risks Ranked as High by EPA Regions.
Source: U.S. EPA 1993, 21.

To summarize, programmatic risk assessments at federal, regional, and local levels have generally produced similar results. These similarities occur despite differences in how the CRAs are conducted as well as differences in real problems that occur in different places. When analyzed in terms of risk instead of public perception, problems such as indoor radon, and indoor air pollution tend to be ranked as high risk. Other problems that tend to rank high in opinion polls are not as frequently identified as posing high risk problems to human health and the environment.

In addition to identifying similar environmental priorities, the programmatic risk assessments conducted to date have helped to identify how CRA methods may be refined. At this writing, EPA's Science Advisory Board is completing a followup study to *Reducing Risk* that is designed to develop better methods to evaluate the costs and benefits of risk reduction options (*Risk Policy Report* 1996b).

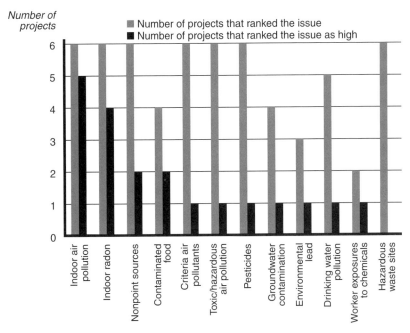

Figure 4. High Priority Human Health Issues Identified by States and Localities.
Source: U.S. EPA 1993, A-3.

Risk rankings are an illuminating way to identify what problems are important to a nation or a jurisdiction. When paired with data that illustrate the cost of pollution control laws, the method becomes a potentially useful way for decisionmakers to assess whether resources are trained on the right problems.

EVALUATING EPA'S PRIORITIES

We have established that programmatic risk assessment paired with a perspective that seeks the most cost-effective risk reduction strategy is a crude but useful tool to inform decisions about how to allocate resources. Given these caveats, what do cost data (presented earlier in this chapter) paired with results from programmatic risk exercises tell us?

Table 3 contrasts the top five human and environmental risks ranked by the CRAs as "high" with the five problems that receive the least amount of money from both EPA and society. In some cases, risks appear to be inversely proportional to expenditures, most notably for problems such as indoor air pollution and indoor radon. The CRAs ranked problems

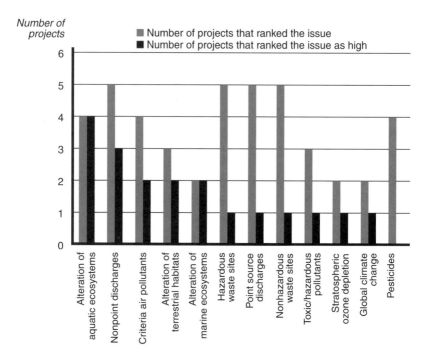

Figure 5. High Priority Ecological Issues Identified by States and Localities.
Source: U.S. EPA 1993, A-7.

such as indoor radon and indoor air pollution as posing the highest risks to human health. While we have no data on how much society spends on these problems, these expenditures are dwarfed by the amount of money society and EPA spend on infrastructure programs to improve surface water quality (see Table 2). Indeed, according to the CRAs, the top risks to ecosystems stem from diffuse sources such as agricultural and urban runoff. A number of EPA regions similarly identified pesticides as posing relatively high risks to people and to ecosystems. Yet pesticides comprise 1 percent of the total cost to society and 2 percent of EPA's expenditures.

Air pollution is the only category where some degree of correspondence appears between spending and relative risk ranking. A number of EPA regions identified hazardous air pollutants as posing high risks to human health. Some states and localities identified criteria air pollutants as posing high risks to both humans and ecosystems. Next to water, society spends most on air pollution. For EPA, air forms the third-highest area of spending, after water quality and Superfund (see Table 2).

Some problems identified by the CRAs, such as the alteration and degradation of terrestrial ecosystems and problems linked to contami-

Table 3. Programmatic Risk Rankings and Expenditures.

Top five programmatic risks				Bottom five expenditures	
EPA regions		State and local programs		EPA	Total society
Humans	Environment	Humans	Environment		
Indoor radon	Terrestrial ecosystem degradation	Indoor air pollution	Alteration of aquatic ecosystems	Indoor radon (0.07%)	Radiation (0.5%)
Other indoor air	Water and wetlands degradation	Indoor radon	Nonpoint sources	Indoor air pollution (0.24%)	Toxic substances (0.8%)
Pesticides	Nonpoint sources	Nonpoint sources	Criteria air pollutants	Radiation (0.26%)	Pesticides (1%)
Hazardous air pollutants	Ozone and carbon monoxide	Contaminated food	Alteration of terrestrial habitats	Leaking underground storage tanks (0.88%)	Leaking underground storage tanks (2%)
Stratospheric ozone depletion	Pesticides	Criteria air pollutants	Alteration of marine ecosystems	Toxic substances (1.2%)	Drinking water (4%)

Source: Table 2, Figures 2 through 5.

nated food, are beyond the immediate scope of the pollution control regulatory system. It is unclear how EPA might address the introduction of exotic species, such as the zebra mussel, that contribute to the alteration of aquatic ecosystems. While the agency has recently stepped up efforts to address problems that stem from nonpoint source pollution, neither Congress nor EPA has made a priority of water pollution problems linked to the runoff of fertilizer from farm fields.

Federal, regional, and state CRAs all fail to identify problems related to hazardous waste as posing high risks to human health and the environment. While society devotes a relatively small amount of money to the clean-up of hazardous waste sites, the Superfund program continues to consume a large portion of EPA's budget.

Most of the aforementioned results relate to human health risks. However, it is much more difficult to evaluate whether and how well the pollution control system is addressing risks to ecosystems, such as habitat deterioration. The difficulty stems in part from the fact that many sources of ecosystem damage are within the jurisdiction of other agencies. For example, factors that contribute to deterioration of habitat arise from such pressures as human population growth, spiraling vehicle use, and encroachment of development into fragile areas. While EPA programs touch tangentially on some of these pressures, such as pollution control requirements for auto air emissions, most of these issues are the responsibility of other agencies and other levels of government. The same applies, at least in part, for some types of nonpoint source problems, such as runoff of fertilizers and pesticides.

Overall, the results of the CRAs, when paired with societal and EPA expenditure data, reveal a mismatch between risks and resource allocation. Unfortunately, the mismatch first identified by *Unfinished Business* and *Reducing Risk*, continues to persist. A 1995 study by the National Academy of Public Administration revealed that while some EPA managers use risk in daily resource allocation decisions, the agency as a whole has not yet achieved a risk-based budget (NAPA 1995). The NAPA report examines in greater detail why it is difficult to reprogram money allocated to EPA each year by Congress from one program area to another. However, beginning in FY1997, Congress restructured EPA's accounts in order to move away from a budgeting process that targets funds to individual media (U.S. OMB 1996). As mentioned, EPA's Science Advisory Board does not expect the results from a followup study to *Reducing Risk* to change much, if at all, from the 1990 findings. In response to the NAPA report and related pressures, EPA in 1996 consolidated its planning and budgeting functions into a new office (*Risk Policy Report* 1996a).

The CRAs have improved our understanding of the environmental challenges that face us. While the pollution control system has helped to improve the quality of air and water, many problems remain and new problems, such as the potential impact of stratospheric ozone depletion, continue to arise. In addition, the exercises have helped to identify where methods employed to conduct programmatic risk assessment and benefit-cost analysis still need to be improved. As efforts to advance the art and science of priority setting continue, the most obvious remaining challenge is to muster the political will to implement the results of such exercises.

Implementation may require modifying laws that conflict or do not work, as well as making difficult institutional changes in the way in which laws are administered. F. Henry Habicht, a former EPA deputy administrator, has stated, "It is clear, then, that in order to re-focus environmental protection programs where they will provide the greatest risk reduction in the most cost-effective manner, Congress must provide legislative relief to EPA to free it from many of its current constraints" (Habicht 1995, 22). Priority setting, like many other crucial pollution control functions, is handicapped by the hodgepodge of national environmental laws.

REFERENCES

Davies, J. Clarence, ed. 1996. *Comparing Environmental Risks: Tools for Setting Priorities*. Washington, D.C.: Resources for the Future.

Finkel, Adam, and Dominic Golding, eds. 1994. *Worst Things First: The Debate Over Risk-Based National Environmental Priorities*. Washington, D.C.: Resources for the Future.

Habicht, F. Henry. 1995. Coming of Age: EPA's Next 25 Years. *Risk Policy Report*. September 22, pp. 21–22.

Hazilla, Michael, and Raymond J. Kopp. 1986. *The Social Cost of Environmental Quality Regulations: A General Equilibrium Analysis*. Discussion Paper QE86-02. Washington, D.C.: Resources for the Future.

Jaffe, Adam B., Steven R. Peterson, Paul R. Portney, and Robert N. Stavins. 1995. Environmental Regulation and the Competitiveness of U.S. Manufacturing: What Does the Evidence Tell Us? *Journal of Economic Literature* 33:132–63.

Landsberg, Hans H., Leonard L. Fischman, and Joseph L. Fisher. 1963. *Resources in America's Future: Patterns of Requirements and Availabilities, 1960–2000*. Washington, D.C.: Johns Hopkins University Press.

Main, Jeremy. 1991. The Big Cleanup Gets It Wrong. *Fortune* 123(10, May 20): 95–96.

Minard, Richard. 1995. CRA and the States. In J. Clarence Davies (ed.), *Comparing Environmental Risks: Tools for Setting Priorities*. Washington, D.C.: Resources for the Future.

NAPA (National Academy of Public Administration). 1995. *Setting Priorities, Getting Results: A New Direction for EPA*. Washington, D.C.: NAPA.

NRC (National Research Council). 1983. *Risk Assessment in the Federal Government: Managing the Process*. Washington, D.C.: National Academy Press.

Risk Policy Report. 1996a. New EPA Planning, Budgeting Office Likely to Boost Risk Issues. March 15, p. 6.

———. 1996b. SAB Followup Will Address "Quality of Life" Issues. March 15, p. 12.

U.S. EPA (Environmental Protection Agency). 1987. *Unfinished Business: A Comparative Assessment of Environmental Problems*. Office of Policy Analysis. Washington, D.C.: U.S. EPA.

———. 1990a. *Environmental Investments: The Cost of a Clean Environment*. Report of the Administrator. Washington, D.C.: U.S. EPA.

———. 1990b. *Reducing Risk: Setting Priorities and Strategies for Environmental Protection*. Science Advisory Board. Washington, D.C.: U.S. EPA.

———. 1993. *State and Local Comparative Risk: Project Rankings Analysis*. Office of Policy, Planning, and Evaluation. Washington, D.C.: U.S. EPA.

———. 1996. Office of Radiation and Indoor Air. Personal communication with the authors, November.

———. 1997. *Summary of the 1998 Budget*. Administration and Resources Management. Washington, D.C.: U.S. EPA.

U.S. OMB (Office of Management and Budget). 1996. *Budget of the United States Government Fiscal Year 1997*. Appendix. Washington, D.C.: U.S. GPO.

U.S. OTA (Office of Technology Assessment). 1995. *Environment Policy Tools: A User's Guide*. Washington, D.C.: U.S. GPO.

7

Efficiency

The Environmental Protection Agency estimates that the United States devotes about $150 billion annually—on the order of 2 percent of the gross domestic product (GDP)—to pollution control. While the nation probably can afford to spend such a sum on environmental protection, it cannot afford to spend such sums unwisely. Certainly expenditures of this magnitude raise questions about the efficiency of our environmental management system. Are we getting good value for our money? Could we do better?

To answer such questions, we turn to the field of economics—and, particularly, to the burgeoning field of environmental economics. Economic analysis can provide an accounting framework for tracking and exploring the implications of environmental decisions. It can array information about the benefits and costs of environmental policies and reveal the cost-effectiveness of alternative approaches. It is, in fact, virtually impossible to insulate major environmental decisions from economic considerations. If economic analysis is not done explicitly, it almost certainly occurs implicitly. In the absence of analysis, decisionmaking is driven by public fears, special interest lobbying, and bureaucratic preferences.

The first part of this chapter addresses the basic question of how much value we are getting for the resources committed to environmental protection. To do this we examine the aggregate or economywide benefits and costs of clean air and clean water policies carried out in the United States over the past several decades. The second part of this chapter examines the cost-effectiveness issue: are we obtaining our environmental goals in an efficient, low-cost manner? Or, can we get more environmental protection for the resources already committed? The third part of the chapter looks at the macroeconomic effects of pollution control. How does the system affect the GDP, employment, and productivity?

ECONOMICS AND ENVIRONMENTAL PROTECTION

Over many centuries, the great transformation from primitive life to civilized society has altered the environments of entire continents, as humans converted forests into farms and domesticated plants and animals. More recently, people's use and adaptation of the environment have increased dramatically with the development of new polluting technologies (such as chlorofluorocarbons). Rising human population and per capita consumption levels have further contributed to the speed at which the environment can be altered. The environmental impacts of civilizations' newly acquired powers are forcing society to recognize that the environment consists of scarce, even exhaustible resources. That is where economics, the science of allocating scarce resources among competing ends, enters. Institutions that were effective when environmental side effects were minimal may be inadequate to deal with these new situations.

The challenge for environmental policy is to determine when public intervention in the affairs of firms and individuals is desirable for environmental reasons, and which policies are most appropriate in various circumstances. Economic analysis can help policymakers identify interventions that generate more benefits than costs and assist them in choosing the intervention that best fits the circumstances.

Economic analysis of environmental programs and policies can serve multiple purposes. It can help allocate resources more efficiently, encourage transparency and accountability in decisionmaking, and provide a framework for consistent data collection and identification of gaps in knowledge. Economic analysis also allows for the aggregation of many dissimilar effects—for example, improvements in health, visibility, and agricultural output—into one measure of net benefits expressed in a single currency.

While societal understanding of the link between the economy and the environment has clearly advanced in recent years, on both the theoretical and the practical side controversy still exists over some of the most basic issues. Some of the controversy stems from the imprecision of the questions themselves (for instance, what do "value" and "cost" really mean in the context of environmental health or natural resources). Other controversy stems from differences in methodology of various empirical studies used to quantify the sometimes-abstract concepts involved in applying economic analysis to the environment.

Environmental Benefits and Costs

Environmental policies can improve human health, increase output of forests and other natural resources, reduce corrosion or soiling of economic assets, and enhance recreational and other environmental assets. A taxonomy incorporating broad categories of environmental benefits is shown in Table 1. Some environmental benefits, such as increased output of forests, are measured in commonly used indicators of economic activ-

Table 1. Taxonomy of Benefits of Environmental Policy.

To individuals	Reduction in mortality
	Reduction in morbidity (acute, chronic)
To production/consumption	Crops/forests/fisheries
	Water-using industries
	Municipal water supply
To economic assets	Reduced corrosion, soiling of materials
	Improved property values
To environmental assets	Recreation
	Other use values (visibility)
	Nonuse (passive use)

ity like the gross domestic product. Other benefits are the nonmarket, welfare-enhancing type that are not represented in the GDP; examples include improved human health or greater biodiversity. Freeman (1982) estimates that more than 90 percent of the environmental benefits of the Clean Air Act and Clean Water Act are of the nonmarket, welfare-enhancing type. Although researchers are trying to develop comprehensive measures of economic activity that capture a broad set of environmental benefits (and costs), theoretical and practical reasons exist for excluding such welfare-enhancing benefits from a commonly used measure of economic activity like GDP. Yet no one doubts that human welfare—rather than GDP—is what societies are ultimately concerned with. The fact that such welfare-enhancing benefits are difficult to value and often involve specialized terminology and measurement techniques does not mean that they are any less valuable than the benefits that are measured by the GDP.

Estimating costs can be as difficult as estimating benefits—a fact not often appreciated by even the most knowledgeable practitioners in the field. For example, the most commonly used measure of environmental costs is reported out-of-pocket expenditures for regulatory compliance. However, this is a narrow measure that may either under- or overstate total compliance costs. On the one hand, the omission of items like legal expenses and diverted management focus suggests that reported out-of-pocket expenditures would tend to understate total compliance costs. On the other hand, failure to account for improved worker health or increased innovation tends to overstate total compliance costs. Table 2 contains a taxonomy of environmental compliance costs. The reader should note, however, that the cost measures reported in this chapter represent reported out-of-pocket expenditures for regulatory compliance. Depending on the importance of the other (generally less well-measured) cost items listed in Table 2, reported compliance costs may under- or overstate total compliance costs.

Table 2. A Taxonomy of Environmental Policy Costs.

Government administration of environmental statutes and regulations	Monitoring Enforcement
Private sector compliance expenditures	Capital Operating
Other direct costs	Legal and other transactional Shifted management focus Disrupted production
Economy-wide effects	Product substitution Discouraged investment Retarded innovation
Transition costs	Unemployment Obsolete capital
Social impacts	Loss of well-paying jobs Economic security impacts
Offsetting benefits	Resource inputs Worker health Increased innovation

Source: Adapted from Jaffee and others 1995, 139.

BENEFITS AND COSTS OF THE CLEAN AIR ACT AND CLEAN WATER ACT

Following this brief introduction to environmental benefits and costs, we may now turn to the first of our two basic questions: what do we know about the overall health, welfare, and ecological and other economic benefits of the Clean Air and Clean Water Acts, and how do these benefits compare with estimated costs? Can we apply an economic framework to determine whether we are getting value for our money?

Early Research

In a major study, Freeman (1982) developed the first and—until quite recently—the only comprehensive benefit-cost analysis of the Clean Air and Clean Water Acts. His study was controversial because it employed a great many assumptions—some would say "leaps of faith"—and attempted to reduce a complex set of issues to a few numbers. One of the key issues in such studies is the question of a baseline: what would air and water conditions have been in the absence of federal legislation? Ideally, one would compare environmental quality levels with and without federally mandated controls while holding all other things constant, including the patterns of production, technology, and demand that determine the generation of pollutants. Such a measure should compare an observed outcome resulting from the policy with a hypothetical or counterfactual position reflecting the same underlying economic conditions and differing only with respect to the impact of the environmental policy. Unfortunately, data and resource limitations prevented Freeman from making such a comparison in his original study. Instead, he measured the benefits of air pollution by examining actual improvements in air quality observed between 1970 and 1978. Water pollution control benefits were measured in terms of expected improvements in water quality between 1972 and 1985. In both cases, as Freeman notes, the measures are likely to underestimate the true benefits because they fail to account for the significant economic growth over the period.

Table 3 presents Freeman's results for air pollution benefits and costs in 1978 (converted to $1995). For stationary sources, total benefits range from $10.1 to $104.2 billion with a most likely point estimate of $45.1 billion. Overall, more than 75 percent of the benefits are health-related. Costs of the stationary source program are estimated to be $19.0 billion. Freeman (1982, 1) concluded that "a fair confidence can be attached to the statement that the control of stationary sources has been worthwhile on benefit-cost grounds."

For mobile sources, Freeman (1982, 131) finds a very different situation. His best estimate of benefits was only $0.6 billion, compared with $16.0 billion of costs. His conclusion was that "the mobile source control program, at least in its present form, does not pass a benefit-cost test."

Table 4 presents Freeman's estimates of benefits and costs for conventional water pollutants in 1985. His most likely point estimate of benefits was $22.0 billion, compared with estimated costs of $36.9 billion (all measured in $1995). He concluded (1982, 170) that "although the upper bound estimate of benefits lie above the range of estimated costs, it is likely that on balance the benefit-cost relationship for the present water pollution control program is unfavorable."

Recent Analyses

Drawing on the sizable body of research conducted in the intervening years, two recent analyses have expanded on Freeman's original work. Like Freeman's estimates, the new studies synthesize and integrate a large body of information derived from the scientific and economics literatures. The first study, an assessment of the benefits and costs of the Clean Air Act from 1970 to 1990, was mandated by Congress as part of the 1990 Clean Air Act Amendments. The study was developed by EPA in conjunction with a congressionally mandated panel of distinguished

Table 3. Air Pollution Control Benefits and Costs in 1978 (in billions of $1995).

Category	Range	Most likely point estimate
Stationary sources		
Benefits		
Health		
Mortality	5.9–58.7	29.3
Morbidity	0.6–26.2	6.5
Soiling and cleaning	2.1–12.7	6.3
Vegetation	0.0–00.0	0.0
Materials	0.8–02.3	1.5
Property values	1.9–14.6	4.9
Total	10.1–104.2	45.1
Cost		19.0
Mobile sources		
Benefits		
Health	0.0–0.8	0.0
Vegetation	0.2–0.8	0.6
Materials	0.8–3.0	1.5
Property values	0.0–4.2	0.0
Total	0.2–3.6	0.6
Cost		16.0

Source: Freeman 1982, 128–30.

Table 4. Benefits and Costs in 1985 From Removal of Conventional Water Pollutants (in billions of $1995).

Category	Range	Most likely point estimate
Benefits		
Recreation	3.8–18.4	9.9
Nonuser benefits	1.1–8.4	2.5
Commercial fisheries	0.8–2.5	1.7
Diversionary uses		
Drinking water: health	0.0–4.2	2.1
Municipal treatment	1.3–2.5	1.9
Households	0.2–1.1	0.6
Industrial supplies	0.8–1.7	1.3
Total	8.0–38.8	19.8
Costs	31.6–42.2	36.9

Source: Freeman 1982, 170.

economists and scientists. The second study was developed in 1994 by ten federal agencies, including EPA and the President's Council of Economic Advisors, in support of a series of proposals by the Clinton administration to augment the Clean Water Act, to include additional nonpoint source controls, more stringent toxics limits, and other provisions.

The Clean Air Act

The new EPA study updates the Freeman methodology by developing and comparing two scenarios as a basis to evaluate progress under the Clean Air Act (CAA): a "no-control" scenario and a "control" scenario. The former scenario essentially freezes federal, state, and local air pollution controls at levels of stringency and effectiveness that prevailed in 1970 and attributes the benefits and costs of all air pollution controls from 1970 to 1990 to federal law.

In all likelihood, state and local regulation would have mandated some air pollution controls even in the absence of the Clean Air Act. States such as California might have required very tight controls. Also, industry likely would have acted on its own to reduce at least some of its emissions. If one assumed that state and local regulations would have been equivalent to federal regulations, a benefit-cost analysis of the federal Clean Air Act would be a meaningless exercise: both benefits and costs would equal zero. On the other hand, any attempt to predict how states' and localities' regulations or voluntary efforts would have differed from the Clean Air Act is extremely speculative. Thus, the freezing of emissions at 1970 emissions rates is a reasonable, albeit unrealistic, assumption.

In the "control" and "no-control" scenarios, the benefits and costs of the act are estimated by a sequence of economic, emissions, air quality, physical effect, economic valuation, and uncertainty models. While each of these steps involves complex scientific and technical issues, the economic valuation step may be the most difficult and is certainly the most disputatious. For details of these models, the reader is directed to the study itself (U.S. EPA 1997).

Table 5 shows selected health benefits, in thousands of cases reduced per year. The midrange estimates of reduced mortality, for example, show that in 1975 the air pollution controls in place reduced premature deaths attributable to airborne particles (PM-10), ozone, sulfur dioxide, and lead by an estimated 20,000 cases. By 1990 the corresponding number of premature deaths avoided stood at 79,000. Similarly, the mid-range estimate for heart attacks avoided rose from 1,000 in 1975 to 18,000 in 1990, largely due to the reduction of lead in the environment.

To develop estimates of economic benefits, it is necessary to translate these physical effects, in this case mostly avoided health damages, into

Table 5. Selected Health Benefits of the CAA, 1970–1990 (in thousands of cases reduced per year, except as noted).

Health effect		1975	1980	1985	1990
Mortality	high	38	97	124	140
(PM-10, O_3, SO_2, Pb)	mid	20	54	70	79
(thousands)	low	11	30	40	45
Heart Attacks	high	1	9	19	24
(Pb)	mid	1	7	14	18
(thousands)	low	1	5	10	13
Strokes	high	1	5	10	13
(Pb)	mid	1	4	8	10
(thousands)	low	1	3	6	7
Respiratory symptoms (SO_2) (thousands)		66	187	165	146
Respiratory illness (NO_2) (millions)		1	4	9	15
Hypertension	high	1	6	12	16
(Pb)	mid	1	5	10	13
(millions)	low	1	4	8	10

Source: U.S. EPA 1997.

dollar terms. As noted, this translation is usually the most contentious aspect of a benefit-cost analysis. Table 6 displays the economic values used in the EPA study. In the case of mortality it is not possible to "value" the lives of victims in a benefit-cost sense. One can, however, determine the compensation required for individuals to accept relatively small reductions in mortality risk. Typically, they are inferred from observed behavior, for example, sales of safety devices such as smoke detectors, or wage differentials associated with high-risk occupations. For expository purposes this valuation is expressed as "dollars per life saved" even though the actual valuation is based on small changes in mortality risk. The estimate of $4.8 million per life saved represents an average value from the literature.

The total monetized economic benefit attributable to the CAA was derived by applying the valuation estimates discussed above to the stream of physical effects calculated for the 1970–1990 period. EPA reports that the estimated benefits of the Clean Air Act realized during the period range from $5.6 to $49.4 trillion, with a central estimate of $22.2 trillion. By comparison, the value of direct compliance expenditures over the same period equals approximately $0.5 trillion. Comparing central estimates, Americans received roughly $45 of value in reduced risks of death, illness, and

Table 6. Central Estimates of Economic Value per Unit of Avoided Effect (in $1990).

Endpoint	Valuation (mean estimate)
Mortality	$4,800,000 per case[a]
Coronary heart disease	$52,000 per case
Chronic bronchitis	$260,000 per case
Strokes[b]	
Males[c]	$200,000 per case
Females[c]	$150,000 per case
Hospital admissions	
All respiratory	$6,100 per case
Ischmic heart disease	$10,300 per case
Congestive heart failure	$8,300 per case
Pneumonia	$7,900 per case
Chronic obstructive pulmonary disease	$8,100 per case
Respiratory illness and symptoms	
Upper respiratory illness	$19 per case
Lower respiratory illness	$12 per case
Acute asthma	$32 per case
Acute bronchitis	$45 per case
Acute respiratory symptoms	$18 per case
Shortness of breath	$5.30 per day
Work loss days	$83 per day
Mild restricted activity days	$38 per day
IQ changes	
Lost IQ points	$3,000 per IQ point
Incidence of IQ < 70	$42,000 per case
Hypertension	$680 per year per case
Welfare benefits	
Decreased worker productivity	$1[d]
Visibility	$14 per unit change
Household soiling	$2.50 per household per PM-10 change
Agriculture (net surplus)	Change in economic surplus

[a]Alternative results, based on assigning a value of $293,000 for each life-year lost are presented (U.S. EPA 1997, ES-9).

[b]Strokes are comprised of atherothrombotic brain infarctions and cerebrovascular accidents; both are estimated to have the same monetary value.

[c]The different valuations for stroke cases reflect differences in lost earnings between males and females. See U.S. EPA 1997, Appendix G for a more complete discussion of valuing reductions in strokes.

[d]Decreased productivity valued as change in daily wages: $1 per worker per 10% decrease in ozone.

Source: U.S. EPA 1997.

other adverse effects for every one dollar spent to control air pollution. As the EPA study notes, "the benefits of the Clean Air Act and associated control programs substantially exceeded costs. Even considering the large number of important uncertainties permeating each step of the analysis, it is extremely unlikely that the converse could be true" (EPA 1997, ES-10).

Clean Water Act

Unlike the new EPA analysis of the Clean Air Act, which focused on total benefits and costs, the interagency study of the Clean Water Act (CWA) estimated *marginal* benefits and *marginal* costs of a set of proposed new initiatives. The baseline assumes current and pending spending as mandated under the 1987 CWA Amendments. (The study also calculated benefits and costs of the new provisions as reductions in cost from stringent interpretations of the 1987 CWA. The report notes (U.S. EPA 1994, ES-2), "while EPA believes that these interpretations are stringent and inflexible, they could potentially be adapted by a court in litigation challenging EPA's interpretation".)

Table 7 presents annualized estimates of aggregate costs from control of urban sources (combined sewer overflow, storm water, toxics) under the initiative and pending spending. Total costs are estimated to range from $10.4 to $14.6 billion. About two-thirds of the costs are borne by municipalities and one-third by private sources. A series of categories is also presented for which costs are not quantified, including those for state administration, federal compliance, and others.

Table 8 presents annualized estimates of aggregate benefits from control of urban toxics under the initiative and pending spending. Six categories of benefits are included: freshwater recreational fishing and swimming, marine recreational fishing, marine nonconsumptive recreation, marine and freshwater commercial fishing, withdrawal or diversionary uses, and human health (exposure via swimming and seafood consumption). Freshwater recreational fishing and swimming benefits account for more than three-fourths of estimated benefits. Assuming no discounting and no lag in attaining water quality benefits, estimated annual benefits range from $860 million to $6.29 billion. Introducing positive discount rates and a lag in attainment of water quality benefits reduces estimated benefits somewhat. Importantly, the study includes a long list of non-quantified benefits, including marine recreational swimming (nonhealth effects), and restoration of biodiversity and ecosystem integrity.

Similar to Freeman's early work, this interagency study finds that costs exceed benefits. However, reflecting the importance of the nonquantified benefits as well as uncertainties in the cost estimates, the authors note: "Despite information [that the quantified benefits are less than the costs],

Table 7. Summary of Aggregate Annualized Costs from Control of Urban
Sources Under the Initiative and Pending Spending (CSOs, Storm Water, Toxics).

Cost category	Range: low–high (millions of $1995)
Municipal costs	
Phase I storm water	$1,730–$2,680
Phase II storm water	$1,080–$2,003
CSO (combined sewer overflow) controls	$3,618
CZARA (nonpoint controls)	$409–$619
Pending spending (Great Lakes, sludge)	$94
Subtotal, municipal costs	$6,931–$9,014
Private sector costs	
Phase I storm water	$2,475–$2,989
Phase II storm water	$362–$1,751
Pollution prevention plans	$63–$126
Domestic sewage exclusion	$294
Nonpoint source controls	$244–$407
Subtotal, private sector costs	$3,438–$5,567
Total quantified costs in urban areas	$10,369–$14,581

Nonquantified costs
State administration costs (urban portion of $682 million)
Federal compliance (urban portion of $991 million)—excludes abandoned mines
Groundwater controls (urban portion of $157 million to $629 million)
Further water quality criteria and standards and sediment criteria
Toxics bans
Other pending spending (such as Great Lakes, pulp and paper effluent
  guidelines, air MACT standards)

Source: U.S. EPA 1994.

the [Clinton] Administration feels it is important to proceed with the Initiative for several reasons [including] …uncertainties [in the cost estimates] and a number of tangible benefits for which monetary estimates have not been developed" (U.S. EPA 1994, ES-10). This initiative is an important application of the notion that quantified benefits (and costs) are not the only effects that can or should have relevance to policy choices. In truth, almost anything can be quantified, but requiring quantification of all effects is an invitation to shoddy analysis and would likely provide no more information—and possibly more sources of dispute and wasted effort—than a careful attempt to consider qualitatively the effects in question. In this instance, the Clinton administration asserted that even though the quantified analysis did not by itself provide a basis to support the initiative, the addition of the nonquantified benefits and uncertainties in the cost estimates were sufficient to proceed with the Clean Water Act proposals.

Table 8. Summary of Aggregate Annualized Benefits from Control of Urban Sources Under the Initiative and Pending Spending (CSOs, Storm Water, Toxics).

Benefit category	Range: low–high (millions of $1995)
Quantified Benefits	
Freshwater recreational fishing and swimming (use and nonuse)	$682–$4,898
Marine recreational fishing (use only)	$42–$461
Marine nonconsumptive recreation (use only)	$31–$315
Marine and freshwater commercial fishing	$42–$199
Withdrawal or diversionary uses	$21–$84
Human health effects (from risks associated with exposure to pollutants via swimming activity and seafood consumption)[a]	$42–$336
Subtotal: quantified benefits[b]	
Assuming immediate attainment of benefits:[c]	
I. Annualized benefits (no lag and no discounting, thus a simple summation of individual categories)	$860–$6,292
Assuming a gradual attainment of benefits over the first 15-year period that all Urban Space Controls are adopted:[d]	
II. Annualized benefits (7% discount rate, gradual attainment over first 15 years)	$587–$4,300
III. Annualized benefits (3% discount rate, gradual attainment over first 15 years)	$692–$5,139
Nonquantified benefits	
Marine recreational swimming (nonhealth effects)	
Other human health effects in marine and freshwaters[a]	
Recreational hunting: freshwater nonconsumptive[a]	
Marine recreational boating	
Other nonuse benefits (marine waters[a])	
Other avoided costs (such as water storage, dredging, damages from floods)	
Restoration of biodiversity and ecosystem integrity	

[a]Given information and methods used to calculate the quantified benefits, some portion of the benefits associated with these categories may be captured in the monetary range ascribed to freshwater recreational fishing and swimming.

[b]Assumes no double counting of benefits or substitution effects between different categories when developing aggregate national estimates. Also assumes that all lower and upper ends of the range for each quantified category describe the aggregate lower and upper bound estimate. Absent information on the distribution or probability of attaining benefits defined by the estimated range, we cannot calculate a "most likely" estimate.

[c]Assuming no lag between implementation of controls, recovery of natural ecological systems, and economic behavior that forms the basis for the economic benefit measures.

[d]These estimates of the economic benefits are more appropriate to use when comparing quantified costs and benefits, given the anticipated lag time between introduction of the control measures and full realization of the environmental and economic benefits. The calculated annualized figure is based on assuming a gradual attainment of benefits up through year fifteen, and a constant future benefits stream after the fifteenth year has been reached.

Source: U.S. EPA 1994.

Table 9. Costs and Benefits of EPA Regulations 1990–1995 (present value in billions of $1995).

	All EPA	*CAA*	*CWA*	*SDWA*	*RCRA*	*CERCLA*
Number of regulations	40	25	3	3	6	2
Costs	187.4	124.7	6.7	25.8	7.8	21.6
Benefits	260.1	214.7	0.5	44.8	0.1	0.0
Net benefits	72.7	90.0	–6.3	19.0	–7.6	–21.6

Source: Hahn 1996.

"Adding Up" Individual Environmental Rules

A somewhat different approach to estimating benefits and costs of environmental regulations was developed by Hahn (1996). Rather than attempt to integrate all clean air or clean water benefits and costs into a single analysis, he examines individual regulations—in this case all major rules promulgated by EPA between 1990 and 1995. Using agency numbers but applying a common discount rate as well as a consistent set of values for reducing health risks, he estimates the present-value benefits and costs from the rules. Table 9 displays his calculations for all EPA rules, as well as estimates for individual programs.

For all EPA regulations, Hahn finds a present-value excess of benefits over costs of $72.7 billion. For the Clean Air Act rules, the present value excess of benefits over costs is $90.0 billion. For the Clean Water Act (based on only three rules), the net excess of costs over benefits is $6.3 billion. Estimates for other programs are also shown. Hahn concludes that "if one takes the Agency numbers at face value there is reason to be gleeful. They basically say that the federal government has done more good than harm in promulgating regulations since 1990" (Hahn 1996, 240). However, he also finds (239) that "about half the final rules would not pass a cost-benefit test."

COST-EFFECTIVENESS OF POLLUTION CONTROL REGULATIONS

Having reviewed a series of large-scale benefit-cost studies to examine the balance between value gained and resources committed to environmental protection, we now turn to a series of distinct but related questions: Is the nation getting as much environmental protection as possible for the resources committed? Could the same environmental benefits be obtained at lower cost? Are the inefficiencies across regulations or regulated parties sufficiently large to suggest that reform opportunities could have substantial payoff?

Cost-effectiveness is a well known analytical technique designed to address such questions. A policy or regulation is said to be cost-effective if it obtains at least cost a particular physical goal, such as tons abated of a particular pollutant. In operation, cost-effectiveness is similar to benefit-cost analysis. The principal advantage of cost-effectiveness is that it avoids the difficult task of finding monetary equivalents for every type of benefit, as was illustrated in Table 6. The principal disadvantage of cost-effectiveness is that the target chosen may be somewhat arbitrary. Over the years, numerous studies have examined the cost-effectiveness of our environmental management system. By defining success according to measures such as pollution abated or human lives saved, most cost-effectiveness studies have measured differences in "bang-for-the-buck" across regulated entities and regulations. Environmental cost-effectiveness studies can be divided into at least three categories: those focusing on differences in pollution abatement costs across producers, those focusing on differences in public health costs across regulations, and those considering alternative policy approaches, such as market mechanisms versus traditional regulatory approaches, in terms of their ability to generate environmental benefits at lower costs.

Abatement Cost Studies

Abatement cost studies are based on the notion that for a pollution control effort to be efficient, the marginal costs of abating pollution should be similar across plants, firms, industries, and sectors. If such costs are grossly unequal, then maximum environmental protection is not being obtained for the amount of money being spent. Consider a hypothetical case of three polluters, each generating ten tons of pollution. Assume that the precise location of the emissions is not critical; that is, there are no "hot spots." (In this example, issues of geographic and demographic distribution of pollution are assumed away. In the real world, however, such issues tend to force policymakers to consider more than just abatement costs in their policy decisions.) The cost for pollution abatement is $10,000 per ton to producer A, $20,000 per ton to producer B, and $30,000 per ton to producer C. If each polluter is required to reduce emissions by one ton, total costs would be $60,000. A more efficient system would allow polluters with higher costs to pay polluters with lower costs to abate more pollution and thereby reduce total costs of obtaining the environmental objective. In this example, the three-ton reduction could be obtained for as little as $30,000. If society deemed it desirable, as much as a six-ton reduction could be obtained for the cost of requiring each producer to reduce emissions by the same (one ton) amount.

Over the years, a fairly extensive literature has developed to address the question of whether abatement costs do, in fact, differ significantly across producers. Some early studies found large differences in cost per unit of pollution avoided among individual plants (Pittman 1979). A review of EPA's process for establishing water pollution control regulations found that application of a uniform cost-effectiveness criterion could have reduced control costs by one-third and at the same time achieved some small additional pollution reduction (Fraas and Munley 1989).

A recent comprehensive abatement cost study estimated the average abatement costs for six pollutant categories for thirty-six different industries from 1979 through 1985. The cost of abating lead emissions, for instance, is estimated to range from $12 per ton in the nonmetal products sector to more than $48,000 per ton in the food sector (Hartman, Wheeler, and Singh 1994). As shown in Table 10, variance in abatement costs for other pollutants was similarly large. If one is confident, as the authors suggest, that the $48,000 estimate is not also capturing other environmental gains, such large variation across industries can be interpreted as a measure of the inefficiency of the regulatory system. That is, disparities are so large that it would probably be possible to achieve the same reductions at considerably lower costs. Alternatively, with the same commitment of resources, greater pollution reduction would be obtainable.

Public Health Studies

Public health studies attempt to detect inefficiencies by examining whether we could be saving more lives at the same cost, or saving the same number of lives at a lower cost. The basic methodology involves estimating differences in the cost of human health protection across regulations. One of the key assumptions of such studies is that the benefits of a regulation can be fully measured in terms of lives or life-years saved. If there are multiple benefits of a regulation, such as reduced morbidity or improved ecological outcomes, the focus on mortality effects may miss some important benefits.

A number of public health studies have revealed large differences in cost-effectiveness across regulations. One study examined the cost-effectiveness of more than 100 environmental interventions in the United States from publicly available economic analyses by providing comparable estimates of each intervention's net resource costs per life or life-year saved (Tengs and others 1995). The authors find that environmental interventions ranged from those, such as the phasedown of lead in gasoline, that have zero or negative net costs to those with much higher costs. For example, they found that certain requirements for chloroform discharges at pulp mills cost almost $100 billion per life-year saved. Another study found the cost-effectiveness of a series of environmental regulations varied dramatically when measured in cost per premature death averted,

Table 10. Average Abatement Cost by Sector, 1979–1985 (in $1996 per ton).

ISI code	Sector	Particulates	Sulfur oxides	NO₂/CO₂	Hydrocarbons	Lead	Hazardous emissions	Other
3110	Food	100	605	266	188	54,066	64	64
3130	Beverages	180	314	13,824	13,824	88	88	88
3140	Tobacco	311	194	194	13,954	149	149	194
3210	Textiles	460	460	1,599	1,599	1,382	1,382	1,382
3211	Spinning	315	621	1,660	218	906	218	218
3220	Apparel	516	71	71	71	71	71	71
3230	Leather	154	438	9,778	734	154	376	495
3240	Footwear	627	240	1,151	1,807	25,682	1,388	1,045
3310	Wood	54	44	44	44	44	44	44
3320	Furniture	49	29	29	29	28	44	29
3410	Paper products	101	422	547	547	101	15,467	101
3411	Pulp, paper	50	180	23	23	40	101	40
3420	Printing	491	136	359	356	136	40	136
3511	Industrial chemicals	53	87	353-	247	1,507	6,500	238
3512	Agricultural chemicals	148	602	1,031	396	617	360	371
3513	Resins	95	652	240	142	504	73	306
3520	Chemical products	246	790	55	183	33	33	237
3522	Drugs	312	1,212	523	200	348	348	348

3530	Refineries	380	191	69	140	3,190	3,190	3,190
3540	Petroleum, coal	69	2,253	90	90	49	49	49
3550	Rubber	254	1,284	398	398	1,172	1,172	1,172
3560	Plastics	254	2,801	272	272	1,313	1,313	1,313
3610	Pottery	215	122	4,398	4,398	5,462	5,462	5,462
3620	Glass	215	638	393	393	215	215	215
3690	Nonmetal products, n.e.c.	23	247	1,911	1,924	13	13	13
3710	Iron, steel	211	612	133	1,395	903	294	21
3720	Nonferrous metals	394	176	57	721	1,489	1,337	205
3810	Metal products	398	1,813	534	463	186	186	495
3820	Other machinery	295	992	597	597	160	160	160
3825	Office, computing machinery	284	284	1,002	1,087	287	20,250	284
3830	Other electrical machinery	433	560	1,808	249	424	192	192
3832	Radio, TV	457	2,151	1,048	1,271	856	1,729	1,320
3840	Transport equipment	737	1,469	543	1,167	543	2,704	543
3841	Shipbuilding	145	965	2,585	2,585	97	97	97
3843	Motor vehicles	406	1,766	1,340	2,831	24,919	184	184
3850	Professional goods	1,401	3,533	1,011	1,596	1,154	1,895	1,154
3900	Other industries	44	30	127	127	30	30	30

Source: Hartman, Wheeler, and Singh 1997.

from hundreds of thousands to trillions of dollars (Hahn 1994; see Table 11). While some qualifications do apply, such large differences in the cost per life saved probably do indicate serious inefficiencies in the overall system designed to obtain public health improvements. Certain studies have examined how much environmental or public health protection we are getting for our money over time, measured either in terms of pollution abated or health protected. In general, they have shown that the cost-effectiveness of environmental regulations has diminished over time, that there are "diminishing returns" to our environmental regulations (Fraas and Munley 1989; Hahn 1994).

Economic Incentives

Looking more broadly at policy tools rather than specific regulations, we can compare the cost-effectiveness of traditional regulatory measures with the cost-effectiveness of other environmental policy tools such as economic incentives. One of the most widely known incentive-based policies, developed pursuant to Title IV of the Clean Air Act, uses a market in tradable "emission allowances" as the way to encourage the electricity industry to minimize the costs of reducing sulfur dioxide emissions. Under this system, the electricity industry is allocated a fixed number of total allowances and firms are required to hold one allowance for each ton of sulfur dioxide they emit. Firms may choose to bank allowances for future use, buy allowances to meet their requirement rather than reduce emissions, or sell excess allowances not required for compliance.

Since the first phase of the program began in January 1995, the experience has been full of pleasant surprises. Firms are reducing emissions in excess of statutory requirements in Phase I and accumulating an unexpectedly large bank of allowances for use in the second phase of the program that begins in January 2000. The experience to date illustrates that the incentives for firms to innovate and to exploit advantageous trends in fuel markets and elsewhere have paid off handsomely. Allowance prices, for example, have declined from EPA's 1990 estimate of $750 per ton to less than $100 per ton in 1997. Although not all of that difference is attributable to efficiency gains arising from the emissions trading program, some experts have estimated that anywhere from one-half to two-thirds of the estimated price decline is attributable to the program (Bohi and Burtraw 1997). Another well-known example of an incentive-based program involves the banking and trading of lead rights instituted in the mid-1980s by EPA as part of the phasedown of lead in gasoline. More than half the nation's oil refineries participated in the program. Savings from the trading program alone are estimated to exceed $200 million. In

Table 11. Cost-Effectiveness of Selected EPA Regulations.

Regulation (70-year lifetime exposure assumed unless otherwise specified)	Year Issued	Cost per premature death averted (millions of $1995)
Trihalomethane drinking water standards	1979	0.2
Cover/move uranium mill tailings (inactive sites)	1983	36.4
Cover/move uranium mill tailings (active sites)	1983	51.7
Standards for radionuclides in uranium mines (45-year lifetime exposure)	1984	3.9
Benzene NESHAP (original: fugitive emissions)	1984	3.9
Arsenic emissions standards for glass plants	1986	15.5
Arsenic/copper NESHAP	1986	26.4
Benzene NESHAP (revised: coke byproducts) (45-year lifetime exposure)	1988	7.0
Hazardous waste land disposal ban (first third)	1988	4,817.2
Municipal solid waste landfill standards (proposed)	1988	21,964.9
Asbestos ban	1989	127.3
Hazardous waste listing for petroleum refining sludge	1990	31.7
Benzene NESHAP (revised: transfer operations)	1990	37.8
Benzene NESHAP (revised: waste operations)	1990	193.4
Hazardous waste listing for wood preserving chemicals	1990	6,552,564.0
Ethylene dibromide drinking water standard	1991	6.6
1,2-dichloropropane drinking water standard	1991	750.7
Atrazine/alachlor drinking water standard	1991	105,840.8

Source: Hahn 1994.

addition, the trading program is credited with accelerating overall program implementation, thereby increasing benefits considerably (Nichols 1997).

Apart from actual outcomes from these and a limited number of other real world examples, an extensive literature examines the *potential* to reduce program costs through use of incentive based policies. Atkinson and Tietenberg (1991), for instance, examined the implications of trading rules in the Cleveland and St. Louis airsheds, finding that the use of trading would represent a substantial improvement over exclusive reliance on the current allocation of control responsibility. Rubin and Kling (1993) find gains from trading, averaging, and banking marketable permit systems would be up to six percent for the market for light-duty vehicles sold in California from 1990 to 2009. Reviewing the literature on incentive mechanisms for environmental management, Carlin (1992) reveals command-and-control to least-cost ratios ranging from 1.5 to 22.0. Table 12 summarizes the potential effectiveness of alternative policy approaches, as estimated in more than two dozen studies.

MACROECONOMIC EFFECTS

Does environmental regulation impede or enhance productivity to the economy as a whole? Is there an inherent trade-off between jobs and the environment? We are all familiar with stories about communities hurt by efforts to protect endangered species, such as the now-infamous spotted owl. Are these stories the exception or the rule? The remainder of this chapter analyzes the macroeconomic consequences of environmental regulation.

Environmental Regulation and Productivity

The United States in 1997 is in its sixth consecutive year of economic expansion. The economic growth rate is higher and unemployment rate lower than those of other major OECD nations. At the same time, we enjoy a reputation as a global environmental leader (see Chapter 9). EPA estimates that we devote about $150 billion annually to environmental protection. As a proportion of GDP—about 2 percent—this amount exceeds that of other nations. This rosy assessment of the mid-1990s differs considerably from the 1970s, however, when the era of national environmental regulation began. At that time productivity growth slowed considerably below the 3 percent growth experienced in the 1950s and 1960s. Many have wondered whether environmental regulation is at least partially responsible for the observed decline in economic growth during the 1970s.

A number of studies have examined the question of whether and to what extent environmental protection hinders growth and productivity. In general, they have found modest adverse impacts from environmental regulation, although some studies have found larger effects and others, smaller ones. Studies using dynamic general equilibrium models—large computer models that include the interactions among all major segments of the economy—have found significant adverse economic effects of environmental regulation (Hazilla and Kopp 1990; Jorgenson and Wilcoxen 1990). Others have failed to find such adverse effects. For example, one study exploring the relationship between state economic growth and environmental regulation, found either insignificant or positive relationships between economic indicators and the strength of state environmental regulation (Meyer 1995, 11). When the analysis covers most or all manufacturing industries, environmental regulations are estimated to account for 8 to 16 percent of the slowdown in the rate of growth (Denison 1979; Gray 1987; Haveman and Christiansen 1981; and Norsworthy, Harper, and Kunze 1979). "Thus," as several noted experts argue, "regulation cannot be considered the primary cause of the productivity slowdown [experienced during the 1970s]" (Jaffe and others 1995).

Higher Cost?

In order to comply with environmental standards, firms must incur certain costs. If they address compliance by adding a cleanup phase to the production process, then their costs are likely to rise. Examples of such measures include catalytic converters on cars or scrubbers on power plant stacks. Investment in pollution abatement equipment imposes significant opportunity costs on firms as it displaces investment in technology that would otherwise enhance productivity. According to a September 14, 1994 article in *The Wall Street Journal*, "over the past 20 years, U.S. businesses have been forced to divert almost 15 percent of their total fixed investment to nonproductive environmental equipment. The total adds up to more than $1.2 trillion" (Bartlett 1994, A18). In addition, firms may encounter indirect costs as a result of price increases of inputs also affected by regulation.

There is an alternative to this ever-escalating cost scenario, however. Environmental regulation can encourage firms to gain competitive advantage by investing in cleaner and more efficient technology. In contrast to the end-of-the-pipe approach described above, many firms, such as Xerox and 3M, have found that a pollution prevention approach that aims to reduce pollution and waste at the input stage can actually improve efficiency (Hart 1997). Unfortunately, some current regulation discourages technological innovation by mandating or strongly encouraging the use of specific technologies (Goodstein 1995, Chap. 14 and 15). There is much interest in devising environmental regulations that avoid this pitfall and take advantage of the "technology-forcing" potential of setting high standards without mandating a specific way to comply with those standards.

Any discussion of the impact of environmental regulation on productivity must consider the health benefits resulting from such regulation (Goodstein 1995). As noted in the new EPA study of benefits and costs of the Clean Air Act, this is no simple task. Given the need to establish a baseline to measure environmental standards that would exist without current regulation in order to extrapolate the impact existing regulation has had on human health, it is quite difficult to conduct credible studies of this sort. Nonetheless, we can infer that, in the absence of environmen-

Table 12. Quantitative Studies of Economic Incentive Savings.

Pollutants controlled	Study	Geographic area	Command-and-control approach	Ratio of CAC cost to least cost
AIR				
Criteria air pollutants				
Hydrocarbons	Maloney and Yandle (1984)	Domestic Dupont plants	Uniform percentage reduction	4.15 (based on 85% reduction of emissions from all source)
Lead in gasoline	U.S. EPA (1985)	United States	Uniform standard for lead in gasoline	(the trading of the lead credits reduced the cost to refiners of the lead phasedown by about $225m)
Nitrogen dioxide	Seskin, et al. (1983)	Chicago	Proposed RACT regulations	14.4
NO_2	Krupnick	Baltimore	Proposed RACT regulations	5.9
Particulates (TSP)	Atkinson and Lewis (1974)	St. Louis	SIP regulation	6.00 (ratio based on 40 g/m³ at worst receptor, as given by Tietenberg, 1985)
TSP	McGartland (1984)	Baltimore	SIP regulations	4.18
TSP	Spofford (1984)	Lower Delaware Valley	Uniform percentage reduction	22.0
TSP	Oates, et al.	Baltimore	Equal proportional treatment	4.0 at 90 ug/m³
Reactive Organic Gases/NO_2	SCAQMD (Spring 1992)	Southern California	Best available control technology	1.5 in 1994
Sulfur dioxide	Roach, et al. (1981)	Four Corners area	SIP regulation	4.25
Sulfur dioxide	Atkinson (1983)	Cleveland		About 1.5
Sulfur dioxide	Spofford (1984)	Lower Delaware Valley	Uniform percentage reduction	1.78
Sulfur dioxide	ICF Resources (1989)	United States	Uniform emissions limit	5.0
Sulfates	Hahn and Noll (1982)	Los Angeles	California emissions standards	1.07 (ratio based on a short-term, one-hour average of 250 g/m³)
Six air pollutants	Kohn (1978)	St. Louis		

Other				
Benzene	Nichols, et al. (1983)	United States	Proposed emission standards	1.96
CFCs	Palmer, et al. (1980); Shapiro and Warhit (1983)	United States	Proposed emission standards	
Airport noise	Harrison (1983)	United States	Mandatory retrofit	1.72 (calculated by Carlin)
WATER				
Biochemical oxygen demand (BOD)	Johnson (1967)	Delaware Estuary	Equal proportional treatment	3.13 at 2mg/l DO; 1.62 at 3mg/l; 1.43 at 4 mg/l
BOD	O'Neil	Lower Fox River, Wisconsin	Equal proportional treatment	2.29 at 2mg/l DO; 1.71 at 4mg/l; 1.45 at 6.2 mg/l
BOD	Eheart, et al. (1983)	Willamette River, Oregon	Equal proportional treatment	1.12 at 4.8 mg/l DO;1.19 at 7.5 mg/l
BOD	Eheart, et al. (1983)	Delaware Estuary in Pennsylvania, Delaware, New Jersey	Equal proportional treatment	3.00 at 3 mg/l DO; 2.92 at 3.6 mg/l
BOD	Eheart, et al. (1983)	Upper Hudson River in New York	Equal proportional treatment	1.54 at 5.1 mg/l; 1.62 at 5.9 mg/l
BOD	Eheart, et al. (1983)	Mohawk River in New York	Equal proportional treatment	1.22 at 6.8 mg/l
Heavy metals	Opaluch and Kashmanian (1985)	Rhode Island jewelry industry	Technology-based standards	1.8
Phosphorus	David, et al. (1977)	Lake Michigan		

Note: The references for the various studies listed in the second column are from Carlin 1992.

Source: Carlin 1992.

tal regulation, higher emissions would lead to a decline in productivity due to increased illness and premature mortality as well as the diversion of investment into medical expenses.

Environmental Regulation and Unemployment

How does environmental regulation affect unemployment? A 1990 poll from *The Wall Street Journal* reported that one-third of Americans thought "it somewhat or very likely that their own job was threatened by environmental regulation" (Rosewicz 1990, A1). A survey by the U.S. Department of Labor suggests that this perception is hugely inflated (U.S. Department of Labor 1987–1990). Based on the experience of 75 percent of all large U.S. employers and 57 percent of manufacturing plants, the survey found that employers attributed only one-tenth of 1 percent of layoffs between 1987 and 1990 to environmental regulation. For the economy as a whole, some studies suggest that environmental regulation has actually resulted in a small net gain in employment (Goodstein 1995). One study finds that roughly 90 percent of the expense of environmental compliance is reinvested in the private sector (Meyer 1995, 11). Meyer points out that instances in which environmental regulation prevents the implementation of specific projects are rare. "Far more often, projects require modifications to meet environmental standards" (Meyer 1995, 14). Goodstein notes that "somewhat surprisingly, environmental protection provides employment heavily weighted in the traditional blue-collar manufacturing, transport, communication, and utility sectors and away from services, both private and governmental" (1995, 117). Thus, while the pollution control regulatory system does occasionally contribute to short-run unemployment in certain pollution-intensive industries, it also provides many employment opportunities.

The rhetoric surrounding the environment/jobs trade-off is best understood in a historical context. Import competition from developing countries, and U.S. investment in manufacturing facilities abroad has contributed to a substantial decline in domestic manufacturing jobs over the past twenty-five years. Some argue that environmental regulation encourages manufacturers to locate in countries with less-stringent regulations (so-called pollution havens). Numerous studies have shown, however, that relocation is driven primarily by cheap labor and access to markets (Bartik 1988). "Even in the most polluting industries, the burden of employee payroll is approximately ten times greater than environmental costs" (Meyer 1995, 12). The effect of labor cost considerations is especially great given the large wage differential between the United States and developing countries. In the late 1980s the average hourly wage in Los Angeles was $8.92, as compared with only $0.77 across the border in Mexico (Goodstein 1995, 121; Grossman and Kreuger 1991). While companies undoubtedly take advantage of lower environmental standards after they relocate, the majority of research suggests that only a very small portion of foreign direct investment is fueled by the desire to avoid tough environmental regulation. For all but the most heavily regulated industries, the cost of complying with environmental regulation is a relatively small fraction of total production costs. Thus, it is not surprising that the employment effects of environmental regulation are also quite small.

CONCLUSIONS

Overall, the benefits of the Clean Air Act since its enactment in 1970 clearly outweigh the costs. Several studies using different methods and different data sources conclude that aggregate benefits exceed aggregate costs. One study finds that benefits exceed costs for rules promulgated from 1990 to 1995. Consistently, the studies show that the health benefits are by far the largest part of the monetized benefits.

The Clean Water Act story is quite different. Three studies show that the costs of the act exceed its benefits. However, there is good reason to believe that important nonquantitative benefits have been omitted from the calculus. As noted in the recent Clinton administration study of the benefits and costs of a proposed initiative, "there are great uncertainties associated with both the cost and benefit estimates that are not captured in the …numerical results…. There are a number of tangible benefits for which monetary estimates have not been developed" (U.S. EPA 1994, ES-9–ES-10). Further research is clearly needed in this area.

Cost-effectiveness analysis reveals large measured differences in costs per ton of pollution abated and in costs per life saved across various regulations. This suggests that considerable opportunities may exist to improve the efficiency of the regulatory system, thereby achieving greater "bang for the buck."

Incentive-based mechanisms, as opposed to traditional regulatory mechanisms, seem to have a great deal of potential to deliver environmental improvements at lower cost. Most recently, the tradable emission allowances for sulfur dioxide have yielded emission reductions at a cost far below what was originally estimated. Other applications of incentive-based mechanisms seem promising.

The macroeconomic effects of pollution control regulation are generally modest. Regulation has had some adverse effect on GDP growth, but most economists think that the effect has been relatively small, and the negative effect fails to take into account most of the benefits of regulation. The effect of environmental regulation on employment is, for all practical purposes, zero. New work by Morgenstern, Pizer, and Shih (1997) indicates that even the reported costs of environmental protection may greatly overstate the true economic effects of regulation.

When looked at as a whole, U.S. environmental progress has made economic sense. It can be shown that benefits exceed costs in a great number of cases. At the same time, it appears as if environmental gains have been achieved at unnecessarily high cost. Thus, the potential economic and environmental gains from reforming the environmental management system are considerable.

REFERENCES

Atkinson, Scott, and Tom Tietenberg. 1991. Market Failure in Incentive-based Regulation: The Case of Emissions Trading. *Journal of Governmental Regulation* 21: 17–32.

Bartik, Timothy J. 1988. The Effects of Environmental Regulation on Business Location in the United States. *Growth Change* 19(3, Summer): 22–44.

Bartlett, Bruce. 1994. The High Cost of Turning Green. *Wall Street Journal*, September 14, p. A18.

Bohi, Douglas R., and Dallas Burtraw. 1997. *SO₂ Allowance Trading: How Experience and Expectations Measure Up*. Discussion Paper 97-24. Washington, D.C.: Resources for the Future.

Carlin, Alan. 1992. *The United States Experience with Economic Incentives to Control Environmental Pollution*. Office of Policy, Planning, and Evaluation. Washington, D.C.: U.S. EPA.

Denison, Edward F. 1979. *Accounting for Slower Economic Growth: The U.S. in the 1970s*. Washington, D.C.: Brookings Institution.

Fraas, Arthur G., and Vincent G. Munley. 1989. Economic Objectives within a Bureaucratic Decision Process: Setting the Pollution Control Requirements under the Clean Water Act. *Journal of Environmental Economics and Management* 17: 35–53.

Freeman, A. Myrick. 1982. *Air and Water Pollution Control: A Benefit-Cost Assessment*. New York: John Wiley.

Goodstein, Eban S. 1995. *Economics and the Environment*. Englewood Cliffs, New Jersey: Prentice-Hall.

Gray, Wayne B. 1987. The Cost of Regulation: OSHA, EPA, and the Productivity Slowdown. *American Economic Review* 77(5): 998–1006.

Grossman, Gene M., and Alan B. Kreuger. 1991. *Environmental Impacts of a North American Free Trade Agreement*. Woodrow Wilson School Discussion Paper 158. Princeton: Princeton University.

Hahn, Robert W. 1994. United States Environmental Policy: Past, Present and Future. *Natural Resources Journal* 34: 305–48.

———. 1996. Regulatory Reform: What Do the Government's Numbers Tell Us? In Robert W. Hahn (ed.), *Risks, Costs and Lives Saved: Getting Better Results From Regulation*. New York: Oxford University Press.

Hart, Stuart. 1997. Beyond Greening: Strategies for a Sustainable World. *Harvard Business Review* 75(1, January–February): 66–76.

Hartman, Raymond S., David Wheeler, and Manjula Singh. 1997. The Cost of Air Pollution Abatement. *Applied Economics* 29(6): 759–74.

Haveman, Robert H. and Gregory B. Christiansen. 1981. Environmental Regulations and Productivity Growth. In Henry M. Peskin, Paul R. Portney, and Allen V. Kneese (eds.), *Environmental Regulation in the U.S. Economy*. Washington, D.C.: Resources for the Future.

Hazilla, Michael, and Raymond J. Kopp. 1990. Social Cost of Environmental Quality Regulations: A General Equilibrium Analysis. *Journal of Political Economy* 98(4, August): 853–73.

Jaffe, Adam B., Steven R. Peterson, Paul R. Portney, and Robert N. Stavins. 1995. Environmental Regulation and the Competitiveness of U.S. Manufacturing: What Does the Evidence Tell Us? *Journal of Economic Literature* 32(March): 132–63.

Jorgenson, Dale W., and Peter J. Wilcoxen. 1990. Environmental Regulation and U.S. Economic Growth. *RAND Journal of Economics*. 21(2, Summer): 314–40.

Meyer, Stephen M. 1995. The Economic Impact of Environmental Regulation. *Journal of Environmental Law and Practice* 3(2, September–October): 4–15.

Morgenstern, Richard D., William A. Pizer, and Jhih-Shyang Shih. 1997. *Are We Overstating the Real Economic Costs of Environmental Protection?* Discussion Paper 97-36-REV. Washington, D.C.: Resources for the Future.

Nichols, Albert. 1997. Lead in Gasoline. In Richard D. Morgenstern (ed.), *Economic Analyses at EPA: Assessing Regulatory Impact*. Washington, D.C.: Resources for the Future.

Norsworthy, J.R., Michael J. Harper, and Kent Kunze. 1979. The Slowdown in Productivity Growth: Analysis of Some Contributing Factors. *Brookings Papers on Economic Activity* 2: 387–421.

Pittman, R.W. 1979. *The Costs of Water Pollution Control to Wisconsin Paper Industry: The Estimation of a Production Frontier*. PhD dissertation. Department of Economics, University of Wisconsin.

Rosewicz, Barbara. 1990. Americans Are Willing to Sacrifice to Reduce Pollution, They Say. *Wall Street Journal*, April 20, pp. A1–A6.

Rubin, Jonathan, and Catherine Kling. 1993. An Emission Saved Is an Emission Earned. *Journal of Environmental Economics and Management* 25(3): 257–74.

Tengs, Tammy O., Miriam E. Adams, Joseph S. Pliskin, Dana Gelb Safran, Goanna E. Siegel, Milton C. Weinstein, and John D. Graham. 1995. Five-Hundred Life-Saving Interventions and Their Cost-Effectiveness. *Risk Analysis* 15(3, June): 369–90.

U.S. Department of Labor. 1987–1990. *Mass Layoffs in* (year). Bureau of Labor Statistics *Bulletins 2395, 2375, 2310.*

U.S. EPA (Environmental Protection Agency). 1994. *President Clinton's Clean Water Initiative: Analysis of Benefits and Costs.* Office of Water. Washington, D.C.: U.S. EPA.

———. 1997. *The Benefits and Costs of the Clean Air Act, 1970 to 1990.* Washington, D.C.: U.S. EPA.

8

Social Values

When people pass judgment on a governmental program, they usually consider the efficiency and effectiveness of the program. They also bring to bear a variety of social values, attributes that society considers important and that, if ignored or violated, can result in fierce opposition to the program in question.

There is no accepted list of such social values. Different types of programs affect different values. Research on how people perceive risk has produced a taxonomy of such values, but risk researchers do not distinguish between personal psychological characteristics (such as fear of the unknown) and social values (such as having a voice in decisions). (See Krimsky and Golding 1992, especially the article by Slovic.)

The list of social values is potentially very long, and the importance accorded to any particular value undoubtedly differs greatly among groups and individuals. Considerable overlap occurs among the different values. We have focused on three sets of values that seem most important for environmental programs: public involvement, environmental justice and equity, and nonintrusiveness.

The importance of social values of the type discussed here cannot be exaggerated. A program could be superbly effective and efficient, but if it seriously failed to meet a widely held social value it would surely be discontinued. In a democratic society, these values are a precondition for successful public policy, and much of the opposition to environmental programs springs from people who believe the programs violate one or more social values.

PUBLIC INVOLVEMENT

Public involvement in the pollution control regulatory system can generally be thought of as interaction between the public and agencies involved in the regulation of pollution. Public education and public participation are ongoing, intertwined processes that may influence one another.

Public involvement in the control of pollution is not limited to inter-actions between government and the public, however. Where industry and other elements of the public mutually interact outside the presence of government, processes of public involvement take shape. In fact, pub-lic involvement in the pollution control regulatory system occurs at many points and in many forms. These include developing federal regulations (rulemaking), environmental permitting, ensuring government and industry accountability, and litigating environmental matters—all of which are considered in this section.

Public involvement in national decisionmaking has been an impor-tant social value since the earliest days of the United States. Over the last quarter century, attention to this social value has grown, reflected by its increasing codification in law. In the case of the pollution control regula-tory system, Congress has frequently incorporated public involvement provisions into environmental legislation.

While public involvement in policymaking is to many an esteemed social value, others consider such involvement, be it for pollution control or other reasons, to be nothing more than pernicious lobbying that enables monied special interests—whether industries, utilities, or envi-ronmental groups—to sway decisionmaking in their favor. In our view, although abuses of participatory mechanisms may occur from time to time, public involvement in the policymaking process is fundamental to the health and vitality of American democracy. Public involvement influ-ences not only the success of a given program but also the public's per-ception of its success. An evaluation of the pollution control regulatory system would therefore be incomplete without consideration of how well that system accommodates public involvement.

Public Involvement As an Evaluative Criterion

Assessing the pollution control regulatory system with respect to public involvement is not a simple endeavor, for several reasons.

First, an evaluation of public involvement lends itself to the subjec-tive biases of the evaluator, since no general agreement exists on what the nature of public involvement in pollution control regulation ought to be. Few individuals fully concur on the desirable degree, character, or scope of public involvement. Some prefer that those who are not directly affected by regulations be totally absent from the process. Others want much greater public involvement.

Then, too, evaluating public involvement entails implicitly or explic-itly defining "the public." Some have argued that each policy issue has its own interested public, including federal regulators, legislators, and inter-ested citizens; others include local opinion leaders, activists, and the citizenry at large (Landy, Roberts, and Thomas 1994, 7–8). The multiplicity of potential "publics" makes it difficult to assess the performance of the system with respect to participation.

Finally, the pollution control regulatory system is fragmented, with different avenues and methods available to interested citizens. By any measure, public involvement varies substantially from EPA program to program, and wide variation exists among states' public involvement efforts. An overall evaluation must recognize and take into account these differences. An evaluation of a single program would not be representative of the system as a whole. With these caveats in mind, we review several processes below to assess both the extent of opportunities they provide for public participation and, more importantly, the degree to which the public influences policy decisions. (Counterparts exist at state and local levels of government to the mechanisms discussed here at the national level.)

Rulemaking and the Notice-and-Comment Process

The development of pollution control regulations provides one of the most important opportunities for public involvement in the pollution control regulatory system. In the United States, the development of federal regulations (or "rulemaking") is carried out through a process known as "notice-and-comment rulemaking." That process, spelled out in the 1946 Administrative Procedure Act, is applicable to most federal government agencies, including EPA. Adequate public involvement in the notice-and-comment process is supposedly ensured through requiring a rulemaking agency to notify the public of a proposed rule and to consider written comments submitted by the public before adopting the rule (Bobertz 1991, 3).

According to the basic notice-and-comment process, a federal agency that has drafted a proposed rule must publish the full text in the *Federal Register*. Also to be included are complete and comprehensible information about the rule (typically a summary of the rule, relevant data, and policy considerations upon which the rule is based) and the processes for finding out additional information and submitting comments (Bobertz 1991, 6–7). Though not required to do so by the Administrative Procedure Act, agencies in some cases give notice of proposed rules through means other than the *Federal Register*, including press releases, public service announcements, and legal notices in newspapers.

By incorporating a notice-and-comment process into the act, Congress intended to provide interested persons with the opportunity to participate in an agency's rulemaking process through the submission of written data, views, or arguments. Though the act does not require public

hearings, it does authorize rulemaking agencies to allow interested parties to communicate their opinions orally on proposed rules. Consequently, some federal agencies, including EPA, hold hearings on proposed rules, enabling interested parties to deliver oral testimony and to
hear the testimony of others. Usually, agencies holding hearings provide
notice of the time and place of the hearings far enough in advance to
allow interested parties to make arrangements to attend and to prepare
testimony (Bobertz 1991, 9).

EPA and other federal agencies are required before adopting a final
rule to "consider the relevant matter presented" by interested parties,
whether in written form or as hearing testimony (Administrative Procedure Act 1946, 5 U.S.C., Section 553: Rulemaking). After such consideration, the rulemaking agency publishes the final rule in the *Federal Register*,
including, in addition to the full text, a statement about the rule's basis
and purpose. To further public understanding of why the rule was
adopted, the published statement may also include a concise summary of
the comments submitted by the public and an explanation concerning
why those comments were either accepted or rejected. Finally, a waiting
period of 30 days is usually observed before the rule is put into effect,
allowing those parties affected by the rule time to adjust their behavior to
the new requirements (Bobertz 1991, 10).

Beyond the basic provisions for public involvement that are built into
the notice-and-comment rulemaking process, additional opportunities
exist for the public to influence rulemaking outcomes. Some of those
opportunities have a basis in law, while others are simply procedures that
rulemaking agencies have adopted. One example of the former is a provi-
sion of the Administrative Procedure Act that concerns the public's right
to petition an agency for rulemaking. Section 553(e) of the act states that
"each agency shall give an interested person the right to petition for the
issuance, amendment, or repeal of a rule."

A second example of a public involvement opportunity sanctioned by
law is that afforded by the activities of "advisory committees" to rulemaking agencies. The Federal Advisory Committee Act of 1972 was enacted to
govern the establishment, operation, and administration of such committees, which the act declared were "a useful and beneficial means of fur-
nishing expert advice, ideas, and diverse opinions to the Federal Government" (Bobertz 1991, 4). The act encourages advisory committees to be
comprised of representatives from industry and citizen organizations,
and it also provides that interested persons be permitted to attend,
appear before, or file statements with advisory committees (Bobertz 1991,
4). As will be discussed later, the act also places limits on the creation and
operation of advisory committees—limits that some observers consider to
be inappropriate.

A third and important example of a public involvement opportunity in the notice-and-comment rulemaking process is "negotiated rulemaking," also known as "regulatory negotiation" (or reg-neg), a procedure that permits an agency to convene a panel of agency representatives and potentially affected parties, including public interest group leaders, to negotiate the text of a rule. Subsequent to negotiations among the parties, the resulting draft rule is subjected to the notice-and-comment process, enabling further public involvement in the rule's development. The steps of the negotiated rulemaking procedure must conform to the requirements of the Federal Advisory Committee Act, as well as those of the Negotiated Rulemaking Act. (Though considered to be an advisory committee, a negotiated rulemaking advisory committee has a narrow mandate, and is usually formed expressly for the purpose of carrying out a regulatory negotiation.) Negotiated rulemaking not only enables the direct involvement of interested parties in rulemaking, but also provides the parties with an opportunity to resolve their differences at an early stage of the rulemaking process, thereby minimizing conflict and legal wrangling over the final rule (Bobertz 1991, 5).

Though not required by statute, certain procedures have been instituted by EPA and other federal agencies to strengthen public involvement in the development of rules. Like negotiated rulemaking, such procedures supplement the notice-and-comment rulemaking process by soliciting public input before an agency has drafted a proposed rule. One procedure, known as Advance Notice of Proposed Rulemaking, provides the public with information about a potential rule, such as the subject matter being reviewed, the reasons why a new rule is called for, options under consideration, and how public comments can be submitted to the agency (Bobertz 1991, 5). These notices are published in the *Federal Register*. Public comment on a notice can be used by an agency in developing a draft rule. In a second procedure, sometimes called "consultation," EPA and other agencies may consult with interested parties to consider their views before drafting a proposed rule. Whether the rulemaking agency convenes all interested parties in a group or solicits the opinion of each sequentially, consultation enables the agency to consider a wide range of viewpoints (Bobertz 1991, 5).

Although the notice-and-comment process as codified presents several opportunities for public involvement in the development of rules, the process has important flaws, in our view: for a given issue, it does not enable the views of the general public to evolve and become defined through interchange, nor does it permit those views to be transmitted in coherent form to agency representatives. The notice-and-comment process has limited ability to foster dialogue between agency representatives and the public. The process is carried out by agencies as much for the sake of appearance as for substantive interchange, and may involve individuals not very representative of the public as a whole. A former EPA general counsel has noted that "notice and comment rulemaking is to public participation as Japanese Kabuki theater is to human passions—a highly stylized process for displaying in a formal way the essence of something which in real life takes place in other venues" (Elliott 1992).

In the last ten years or so, public involvement in the notice-and-comment process has been enhanced by the mechanisms of advance notice of proposed rulemaking, negotiated rulemaking, and consultation. EPA has only infrequently employed two of these public involvement tools in recent years, having issued only eleven advanced notices since April 1994 (Table 1) and having initiated not a single regulatory negotiation during the same time period (Table 2). Though other federal regulatory agencies have not outdone EPA in absolute terms in their use of these mechanisms (as the same tables show), EPA does not compare very well in relative terms. For the time period examined in the tables, when advance-notice and reg-neg numbers for selected agencies are compared with new rulemaking activity (Table 3), only a small fraction of EPA's proposed rules are seen to have made use of these two mechanisms. Prior to April 1994, EPA had been the most frequent user of the negotiated rulemaking mechanism, having initiated 40 percent of all negotiated rulemakings (see Table 2). Since that time, however, EPA has not initiated any regulatory negotiations, although other agencies have continued to use the mechanism.

Table 1. Advance Notice of Proposed Rulemaking: Selected Agencies.

Agency	April 1994–April 1995	April 1995–April 1996	April 1996–April 1997	Three-year total
Environmental Protection Agency	2	1	8	11
Nuclear Regulatory Commission	2	0	2	4
Occupational Safety and Health Administration	0	0	0	0
National Highway Traffic Safety Administration	1	2	1	4
Consumer Product Safety Commission	4	0	3	7
Food and Drug Administration	2	1	10	13
Total for above agencies	11	4	24	39
All federal agency total	58	56	86	200

Source: Federal Register, 1994–1997.

Table 2. Negotiated Rulemakings Initiated, 1994–1997: Selected Agencies.

Agency	April 1994– April 1995	April 1995– April 1996	April 1996– April 1997	Three-year total	Pre-April 1994	Total reg-negs
Environmental Protection Agency	0	0	0	0	16	16
Nuclear Regulatory Commission	0	0	0	0	2	2
Occupational Safety and Health Administration	1	0	1	2	2	4
National Highway Traffic Safety Administration	0	1	0	1	1	2
Consumer Product Safety Commission	0	0	0	0	0	0
Food and Drug Administration	0	0	0	0	0	0
Total for above agencies	1	1	1	3	21	24
All federal agency total	14	8	5	27	41	68

Sources: Administrative Conference of the United States 1995 and the online *Federal Register*. Dates based on issuance of notice of intent to establish negotiated rulemaking committee.

Table 3. Number of Entries into "Proposed Rule" Stage for Selected Agencies.

Agency	April 1994– April 1995	April 1995– April 1996	April 1996– April 1997	Three-Year Total
Environmental Protection Agency	89	115	87	291
Nuclear Regulatory Commission	31	9	7	47
Occupational Safety and Health Administration	1	5	8	14
National Highway Traffic Safety Administration	22	34	28	84
Consumer Product Safety Commission	6	0	1	7
Food and Drug Administration	13	30	19	62
Total for above agencies	162	193	150	505
All federal agency total	NA	NA	NA	NA

Notes: NA indicates that the relevant information was not acquired because of time constraints. "Proposed Rule Stage" *entries* consist of actions for which agencies plan to publish a Notice of Proposed Rulemaking (NPRM) as the next step in their rulemaking process.

Source: U.S. GSA 1994–1997

One explanation for EPA's limited use of advance notices and reg-negs during this period might be its greater reliance on consultation and other forms of "consensus-based mediation" (which comprises a range of mechanisms that encourage resolution of multiparty problems through mediation leading to consensus). EPA's limited use of regulatory negotiations and advance notices results from an increased emphasis, particularly since 1993, on initiatives designed to overcome the cumbersomeness of the existing statutory system. Several of these initiatives, notably the Common Sense Initiative and "Community-Based Environmental Protection," have as a major purpose the facilitating of public participation. In general, they have succeeded in encouraging exchanges of views among stakeholders; they have been less successful, however, in fostering communication between government and interested parties. A most important mechanism for facilitating dialogue between EPA officials and the public is the advisory committee. Advisory committees are a continuing and meaningful method for citizen input and interchange on particular matters, whether these concern the need for a regulation or more general agency policy. As noted above, such interchange is the key ingredient lacking in the standard notice-and-comment form of rulemaking.

Prior to the establishment of the Federal Advisory Committee Act in 1971, some 1,800 advisory committees were handling a multitude of issues of concern to federal agencies (Cardozo 1981, 5). The act set new ground rules for advisory committees, including the requirement that agencies receive permission from the Office of Management and Budget to establish any new committee. By requiring this permission, the act provided leverage for those who thought advisory committees were bureaucratic excess. Spending-conscious administrations, both Republican and Democratic, have periodically fought against creation of new advisory committees, whose staff and operations incur costs. Moreover, presidents (including President Clinton) seeking easy targets for budget reductions have set out to reduce the number of advisory committees. These policies have resulted in nominal changes: in EPA's case, the committee count has fallen from thirty-one in FY1993 to twenty-two as of July 1995 (BNA 1996). At $9 million, EPA's 1995 expenditures on advisory committees were one-seventh of one percent of the total agency budget. The overall number of federal committees is deceptive because, in response to the act's requirements, agencies have created subcommittees of existing committees instead of creating new committees.

In sum, although the notice-and-comment rulemaking process provides several opportunities for public involvement, the process does not sufficiently define the views of the general public on proposed regulations, nor does it enable the public to communicate those views effectively to agency representatives. Furthermore, the notice-and-comment process cannot compensate for the differential access and resources of competing interests, factors that also impede unbiased EPA dialogue with the public (Kraft 1996, 111). Moreover, the provisions of the Federal Advisory Committee Act that limit the use of advisory committees—perhaps the most effective means of interfacing between the public and federal agencies—impede more extensive dialogue between EPA and the public. In recent years, EPA has made only minimal use of advanced notices and negotiated rulemakings, techniques that could improve public involvement.

Environmental Permitting

Permitting provides another opportunity for people to participate in the decisionmaking process. An environmental permit can address the conditions under which a facility may operate, the types and amounts of pollutants it may discharge, and the facility's requirements concerning operation, maintenance, reporting, recordkeeping, and all aspects of monitoring and inspection. The possibility of losing its permit gives a company a powerful incentive to comply with environmental obligations (Bobertz 1991, 11).

State environmental agencies issue environmental permits in order to regulate numerous forms of polluting and potentially polluting enterprises, including facilities generating air and water pollution, mining operations, and operations involving the treatment, storage, and disposal of hazardous wastes. (The federal statutes requiring permits for the above enterprises are, respectively, the Clean Air Act, the Clean Water Act, the Surface Mining Control and Reclamation Act, and the Resources Conservation and Recovery Act.) Each of the permitting programs relevant to these different enterprises has a set of rules for the issuance, modification, and revocation of permits, as well as procedures for public involvement in the permitting process (Bobertz 1991, 12). Agencies notify the public when a permit application is submitted and offer the opportunity for public comment on the application.

Local communities are generally the most likely to be affected by permits granted to facilities, and agencies typically place notices in local newspapers and with radio stations. A permit-granting agency may also send out notices to such potentially interested organizations as local civic associations, local chapters of environmental organizations, trade unions, and recreational organizations (Bobertz 1991, 15). As in the notice-and-comment rulemaking process, agencies allow hearings in the permitting process whenever the public exhibits strong interest.

Upon granting a permit, an agency issues a notice summarizing comments received from the public and testimony presented at hearings; it also explains how public input influenced or failed to influence the final form of the permit. Procedures exist for further administrative or judicial review of an environmental permitting decision, should it be challenged by members of the public or by a permit-seeking facility.

Public involvement in notice-and-comment rulemaking and in environmental permitting has similar faults. Though permitting offers several opportunities for public involvement, outside actors seldom influence outcomes. Also, little opportunity is provided for interchange among regulatory agencies, permit applicants, and the public (see U.S. EPA 1995b, 27). Communities often lack sufficient expertise to understand technical details contained in permits. Consequently, in many cases few or no citizens comment on permit applications. Moreover, community organizations often lack the resources needed to travel to regional offices to review information and records. Agencies may also fail to employ effective notification mechanisms for informing local communities of a permit application, and only rarely do they devote sufficient resources to help citizens understand a permit application. Great variability exists among states in terms of the importance and resources granted by their agencies to public involvement efforts in permitting.

In sum, the environmental permitting process generally does not adequately accommodate public involvement. Both EPA and state agencies need to make greater efforts to engage the public early on in the permitting process and to better educate the public about issues relevant to permits. Adopting appropriate strategies to address these inadequacies might help develop an enhanced relationship and improved dialogue among stakeholders, permittees, and permit writers. EPA's Permits Improvement Team has recently made several promising recommendations that, if adopted, could help to realize these goals (see U.S. EPA 1996, 43–59).

Ensuring Accountability

Citizens play an important role in holding federal agencies and industries accountable for pollution control, serving as a watchdog to ensure that those entities do not violate federal regulations and laws. The public also holds federal and state agencies accountable for their regulatory and enforcement responsibilities. Important laws and their implementing programs that assist such involvement include the Freedom of Information Act, the Emergency Planning and Community Right to Know Act, and the National Environmental Policy Act. These statutes grant public access to information compiled by federal agencies and, in the case of EPCRA, industries. They also allow citizens to sue to correct violations of the law.

The Freedom of Information Act, enacted in 1966 and amended in 1974 and 1986, provides the public a statutory right of access to federal government records (U.S. DOE 1996, 108). It requires that all federal agencies, including EPA, publish in the *Federal Register* basic instructions concerning how and where the public may obtain information about an agency and its activities. The act also requires federal agencies to provide to the public requested documentation and information within a specified period of time and for a reasonable fee, as long as release of the requested records and information does not conflict with certain goals (such as national security, commercial and trade secrecy, and personal privacy).

By guaranteeing public access to government records, the act has abetted public involvement in the pollution control regulatory process and has pressed agencies to be more accountable for their actions. Recently, it has enabled members of the public to access the enormous inventory of records documenting the Department of Energy's polluting activities and their consequences, thereby increasing pressure on the department to meet fully its pollution control requirements. Despite the successes of the act, several federal agencies have historically carried a substantial backlog of information requests, unable or unwilling to meet response deadlines specified by the act.

The Emergency Planning and Community Right to Know Act (Title III of the Superfund Amendments and Reauthorization Act of 1986) established the Toxics Release Inventory (TRI), a computerized national database of toxic chemical releases by individual manufacturing facilities. This publicly accessible database, the first such database ever required under federal law, contains national information at both chemical- and facility-specific levels (Gottlieb 1995, 131–32). Regional information is also available in the database, which can be accessed on the Internet. Using this information, EPA publishes an annual report on the total volumes of toxics released in the United States (see Chapter 5).

The TRI was in part a response to the growing belief that communities have a right to know about manufacturing plants, chemical facilities, and other such industries that might pose hazards (BNA 1988, 792). The consequences of large industrial accidents in Bhopal, India and Chernobyl, Ukraine on local communities contributed to the momentum to develop the TRI. Advocates held that industry would be held accountable for its toxic pollutant releases if information about those releases were available to the public.

The Toxic Release Inventory has empowered the public to influence industry activities involving toxic chemicals. Citizen groups armed with information on the release of toxic chemicals by industry have in some cases been able to pressure industry into reducing such releases. Environmental advocacy groups provide TRI information to local media, and activists have been responsible for fomenting citizen protests, engaging state environmental regulators, initiating state legislative activity, and inducing substantial voluntary compliance by industry (Vallely 1993, 256–57). Thus, the TRI is an accountability mechanism that has contributed to increased public involvement in the pollution control regulatory system while also contributing to pollution reduction.

Enacted in 1970, the National Environmental Policy Act (NEPA) requires that most federal agencies take into account the environmental impacts of their major actions. (EPA is the notable exception). To ensure that agencies adequately carry out this mandate, NEPA obligates them to issue for public scrutiny documentation specifying the potential impacts of major proposed activities. For activities expected to have substantial consequences, agencies must produce an environmental impact statement and make it available to the public. For activities expected to have less than substantial consequences, a more limited and concise document known as an environmental assessment must be produced. Guidelines developed by the Council on Environmental Quality, a federal agency created by NEPA, enable agencies to define major proposed activities and to determine which of the two documents should be produced for any given proposed action.

The environmental impact statement (EIS) process may include the following steps. First, an agency holds a "scoping meeting" at which representatives of interested parties, including businesses and public interest groups, can discuss the key issues that the environmental impact statement should address. The goal of the scoping meeting is to reduce future conflict, litigation, and wasted effort that might result were an EIS later judged to be inadequate. Taking into account the issues raised at the scoping meeting, a draft EIS is then developed and distributed to relevant federal and state agencies, citizen organizations, and other interested parties. These entities submit their comments on the draft to the issuing agency. The draft and final EIS must include a discussion of the current environmental situation; a description of the project proposed by the agency generating the EIS; an analysis of the probable environmental consequences of all phases of the project, including short- and long-term effects; a comparison with feasible alternative options; and suggestions for minimizing adverse effects (Barbour 1980, 189). After receiving comments on the draft EIS from federal, state, and citizen entities, the issuing agency is required to respond in the final EIS to the submitted comments. Though the agency is not required by law to adjust its plans according to submitted comments, changes in the final EIS may reflect the influence of those comments.

NEPA's environmental impact statement process, though imperfect, is a significant vehicle for public involvement through public education and participation and agency disclosure and accountability. It has enabled unprecedented public access to a vast amount of planning information that otherwise would have remained hidden in agency files, while also providing access to the decisionmaking process itself through the possibility of later court action (Barbour 1980, 189). Specifically, although the public cannot legally challenge federal agencies on the appropriateness of the alternatives selected in environmental impact statements, it can challenge the adequacy of an agency's preparation of an EIS. In fact, legal challenges to the adequacy of environmental impact statements have on numerous occasions forced federal agencies to alter or cancel specific plans and projects.

In recent years, some have claimed that the EIS process has become less effective as an instrument of accountability, in part due to changing perceptions in the courts of what constitutes an adequate EIS. The number of legal challenges to environmental impact statements has fallen substantially since the late 1970s as environmental groups have focused their limited resources elsewhere. The *percentage* of EISs challenged in court has remained relatively stable, however, fluctuating between 15 and 20 percent of all EISs filed (Table 4). This trend seems to indicate that the EIS process, despite its flaws, continues to be an important mechanism for drawing public attention to agency activities while forcing agencies to consider the environmental consequences of their actions.

In sum, public involvement mechanisms that promote government and industry accountability for pollution control have proved quite suc-

Table 4. Filings and Challenges of Environmental Impact Statements.

Year	EISs filed	EISs challenged	Percent challenged
1979	1,273	139	10.9
1980	966	140	14.5
1981	1,033	114	11.0
1982	808	17	2.1
1983	677	146	21.6
1984	577	89	15.4
1985	549	77	14.0
1986	521	71	13.6
1987	455	69	15.2
1988	430	91	21.2
1989	370	57	15.4
1990	477	85	17.8
1991	456	94	20.6
1992	512	81	15.8
1993	465	89	19.1
1994	532	106	19.9

Source: U.S. CEQ 1993, 350, 368–69; U.S. CEQ 1997, 534, 543–44

cessful, forcing or encouraging government and industry to alter behaviors or plans identified by the public as not meeting environmental protection goals. The three mechanisms identified above could certainly be improved, but they have served their basic purpose well.

Environmental Litigation

Litigation is another important mechanism for public involvement. Litigation over pollution control is targeted at either industry or government, typically to halt or prevent polluting activities, to require federal agencies to take actions mandated by law, or to challenge agency actions (Bowman 1992, 10). Several aspects of the law affect the use of litigation by citizens to influence environmental policy.

Citizen Standing. To bring a lawsuit to court, a citizen must have "standing," permission as a person with a direct stake in the outcome of a dispute to invoke the jurisdiction of a court to resolve it. Plaintiffs who claim standing have been required to demonstrate that they have been or will be physically or economically injured by the actions or threatened actions of the defendant, that the injury is traceable to the challenged action, and that the harm alleged is likely to be redressed by a favorable decision (Bowman 1992, 28). Prior to the early 1970s, restrictive views of these "standing" requirements limited the scope of citizen-initiated lawsuits. With the advent of the environmental movement and passage in Congress of numerous environmental statutes containing broad citizen-suit provisions, requirements for standing began to evolve. The liberal courts of the time no longer regarded physical or economic injury to be strictly necessary for a plaintiff to have standing. A more flexible notion of standing came to be accepted, one by which certain citizens (plaintiffs with injuries in the "zone of interests" of the statute in question) had the right to initiate a lawsuit even for an aesthetic injury, such as visual impairment in a national park due to emissions from a nearby power plant. This more flexible notion of standing was a key factor in enabling the surge of environmental litigation in the 1970s. More recently, however, increasingly conservative courts have interpreted the notion of standing more strictly, leading some observers to predict that environmental groups will experience diminished access to the courts. Whether a restrictive interpretation of standing is more an impediment or an inconvenience to citizen-initiated environmental litigation is subject to considerable debate. Environmental petitioners will nevertheless have to be more careful in the way they frame their claims to standing (see, for example, McElfish 1993).

Judicial Review. To ensure that federal agencies carry out the tasks assigned to them by Congress, certain statutes allow citizens to seek judicial review of an agency's failure to act as mandated by the statute. After being notified of a citizen's intention to bring suit, a targeted agency that does not soon perform the relevant required action can be taken to court for failure to perform a duty that is "nondiscretionary." In court, the plaintiff need only establish the existence of a statutory duty and the agency's failure to perform that duty. A ruling in favor of the plaintiff may lead the court to order agency performance of the act or duty, along with the establishment of compliance schedules (Bowman 1992, 11).

According to provisions of certain environmental statutes and the Administrative Procedures Act, citizens may also challenge in court the final actions of federal agencies. Such challenges typically concern agency decisions about whether or how to regulate. Although courts typically grant great deference to agencies' interpretation of their own statutory authority, they watch very carefully whether agencies conform to the rigor of the Administrative Procedures Act in the rulemaking process. Generally, an agency's final action will only be rejected by the court if the decision is found to lack any reasonable basis in fact—if it is "arbitrary and capricious"—or if the decision is considered to be "contrary to law." A court may also reject final decisionmaking on the basis of defects in the agency's actions in reaching the decision, such as a determination that an agency did not consider and respond to all comments raised by the public during the rulemaking process.

Many agency final actions involving the environment are litigated. Lawsuits have been filed against federal agencies by both environmental groups and business interests on the grounds that agency-issued regulations were too harsh or too lenient. In addition to suits brought under the terms of the Administrative Procedures Act, lawsuits challenging final actions have often been initiated pursuant to the provisions of the National Environmental Policy Act, with suit brought against federal agencies either refusing to develop an environmental impact statement or developing an inadequate EIS.

Citizen Suit Provisions. In certain situations, federal agencies may choose not to take enforcement actions against violators of environmental law. When federal enforcement is not pursued, citizens may in some cases directly enforce environmental requirements by means of citizen suit provisions, which are contained in most environmental protection statutes. By granting citizens the power of "private attorney general," citizen suit provisions in environmental statutes allow any "person" (including an individual, organization, or corporation) to sue any other "person" violating the requirements of the statute in question (Bowman 1992, 15). Citizen suit provisions extend and supplement the government's enforcement capacity: state and federal agencies do not have adequate resources to bring every case to court.

After a violator has been notified of an impending citizen suit, it is allotted a limited period of time to change its behavior or negotiate with the suing party. (The waiting period also allows for state or federal agencies to prosecute the violations, if they prefer to do so themselves.) If violations continue and the citizen suit is taken to court, the defendant risks being ordered by the court to cease its violations, to recompense the plaintiff for court costs and attorney fees, and (under the Clean Water Act, the Clean Air Act, and the Resource Conservation and Recovery Act) to pay civil penalties for its violations to the U.S. Treasury (Bowman 1992, 15). However, under citizen suit provisions of environmental statutes, plaintiffs are not able to recover any personal damages for violations of environmental laws and regulations.

Common Law Remedies. Members of the public can also bring actions to redress environmental harm using common law remedies. This type of legal action allows plaintiffs to seek damages for harm caused to them by the actions of others. Despite the growth of lawsuits based on citizen suit provisions built into the major environmental statutes, common law remedies still play a role, albeit minor, in the realization of environmental goals. The threat of a substantial damage award arising from a common law remedy, which might include punitive as well as compensatory damages, provides citizens with some ability to deter potentially polluting activities and force industry to pay attention to citizen claims (Bowman 1992, 16).

Under a common law remedy, damages may be sought on the grounds that a defendant's actions caused injury to the plaintiff, posed a nuisance to the plaintiff, or resulted in trespass of the plaintiff's property. In addition, some litigation can be brought as a "class action" on behalf of others similarly situated, to provide a broad-based set of remedies for the harm being litigated. Thus, a property owner annoyed by fumes emitted from an adjacent factory, frustrated by the seepage of oil into his fishpond from the factory's fuel storage facility, and physically ill due to the consumption of well-water contaminated with factory-emitted toxins would have three different reasons to initiate a common law remedy. Were he to be joined in the lawsuit by affected neighbors, the result would be a "class action" lawsuit against the factory owners. (If victorious, a plaintiff may also obtain a court order requiring a defendant to halt damaging behavior.)

Participation in Government Enforcement Actions. In some cases, citizens may "intervene" in a lawsuit brought by the government against an alleged violator of pollution control requirements by joining the lawsuit as an "interested party." The status of interested party enables citizens to take part in negotiations and make their perspective(s) known to the presiding judge, although government prosecutors retain the authority for prosecuting the case. Most U.S. environmental statutes that authorize citizen suits also grant citizens the right to intervene in government enforcement proceedings when certain criteria are met by the interested citizens. A citizen may also participate in a case by filing an "amicus," or friend-of-the-court, brief setting forth his or her position (Bowman 1992, 9). Though court permission is required to file an amicus brief, such permission is generally granted.

Environmental litigation has been a very effective form of public involvement in the pollution control regulatory system, despite its flaws. The advent of environmental litigation in the 1970s provided citizens with a tool to ensure that federal agencies obey and enforce pollution control laws passed by Congress. Since that time, it has acted as a check on agency decisionmaking and assured the fairness of decisionmaking processes. Environmental litigation also strengthened public involvement in rulemaking, permitting, ensuring accountability, and other areas because inadequate attention to these forms of involvement can be redressed in court. Early litigation in federal courts over the enforcement of environmental laws also enabled public interest groups to have demanding interpretations of the laws accepted (Vig and Kraft 1994, 149).

From the perspective of public involvement, environmental litigation also has numerous problems. One is uncertainty concerning the future of citizen standing before the courts. If an increasingly restricted notion of standing were to evolve in the courts, the effectiveness of litigation as a means to advance pollution control could be undermined. Irrespective of standing considerations, the receptivity of courts to arguments for stronger environmental controls or more liberal damage awards will similarly influence whether litigation serves as a useful form of public involvement.

Additional disadvantages of environmental litigation as a mechanism of public involvement can include inefficient use of taxpayer dollars in the form of high government legal expenses, lengthy delays in the implementation of laws, prolongation of environmental problems, overtaxed judicial resources, and the creation of an atmosphere of adversarial hostility that may discourage future cooperation between litigants (Bowman 1992, 19). Litigation can result in the setting of agency priorities by external parties and courts rather than by agency officials or Congress. Finally, since litigation is rather expensive and can be drawn out, those litigants with superior legal and economic resources are likely to fare better than those without such resources.

Other Opportunities for Public Involvement

For many, recycling is the most evident form of public involvement in the pollution control regulatory system. Many communities across the United States either require or encourage community residents to recycle various materials, with the goal of minimizing the amount of trash that must be incinerated or disposed of in landfills. By relying on continuous public participation, recycling presents citizens with a daily lesson about who is ultimately responsible for pollution control—citizens themselves. Though some analysts question the utility of large-scale recycling from the perspective of pollution control, recycling efforts nonetheless instill in citizens awareness of, and concern for, the environment.

Less direct forms of public involvement in pollution control are numerous and quite varied in nature. Information exchange is an example of less direct public involvement, whereby members of the public exchange information with federal, state, and local agencies, with environmental organizations, or with relevant legislators at the state and national levels. Such exchange can consist of one-way or two-way information flows via telephone calls, faxes, letters, meetings, the media, agency publications, hearings, speeches, opinion polls, and the like. It can also occur via electronic mail and the World Wide Web, where the availability of environmental data has expanded dramatically in recent years. Through the exchange of information, citizens may learn additional details about the pollution control regulatory system (public education) or may express their concerns about the system and the problems it attempts to address (public participation); education and participation may occur simultaneously.

Information exchange as a mechanism of public involvement in the pollution control regulatory system has a mixed record. Although many avenues exist for agencies and the public to transmit information, the opportunity for two-way information exchanges is rather limited. In general, beyond activities of the limited number of advisory committees and certain other meetings with the public (such as reg-neg meetings and consultations for rulemaking), most information exchanges between EPA and the public are unidirectional.

According to some, the general public's limited awareness and understanding of the issues that concern EPA, particularly the tradeoffs between health and cost, are an additional failure of EPA's information exchange efforts. Nor have EPA efforts contributed greatly to a more sophisticated public understanding of environmental matters, such as the relative risks posed by different environmental hazards (Landy, Roberts, and Thomas 1994, 311). EPA's growing use of the Internet and other nontraditional mechanisms of public involvement, however, may help to improve public awareness and understanding.

Citizen monitoring, information gathering, and reporting are additional, less direct forms of public involvement. At the local level, these activities entail citizen observation of pollution-related occurrences in a community —such as oil sheen on a river, a smokestack emission, or the activity of a developer in a swamp. Other indirect forms of participation include the purchasing decisions that citizens make each day. By boycotting the products of polluting companies or purchasing "green" products, members of the public may influence the behavior of companies. Also, voting and donations to electoral campaigns might be considered indirect forms of public involvement in the pollution control regulatory system.

NONINTRUSIVENESS

Along with public involvement, nonintrusiveness is an important social value that Americans consider when sizing up the success of a government program. In fact, since the earliest days of the nation's existence, Americans have held dear the principle of minimal governmental interference in the lives of citizens—perhaps the most basic notion of governmental nonintrusiveness. Nonintrusiveness is embodied in the Bill of Rights, which specifies numerous freedoms of which government may not deprive citizens. However, unlike public involvement, nonintrusiveness has not been highly codified into American law The social value of nonintrusiveness is represented in certain statutes. For example, the Paperwork Reduction Act and its various amendments attempt to minimize the burden of paperwork that government agencies place on businesses and state and local governments. The Freedom of Information Act requires that certain government documents remain private if their release were to result in unnecessary privacy violations of citizens. The vigorous antigovernment movements that have sprouted at the same time as the growth of governmental regulation are indicative of the sustained and increasing importance of nonintrusiveness as a social value.

The relevance of nonintrusiveness as a social value varies among individuals, many of whom believe other social values to be more important. For example, a devoted environmentalist might not be at all concerned with the intrusiveness of pollution control programs as long as those programs successfully prevent or mitigate environmental degradation. Conversely, a libertarian might care about the environment yet seek to dismantle pollution control regulations because of their intrusiveness.

More importantly, the perception of what intrusiveness actually is differs among individuals according to factors such as ideological perspective, values, and experience. The libertarian, for example, might consider the smallest constraint imposed by a governmental program to be an intrusion, whereas the liberal democrat might perceive such a constraint to be a normal and acceptable consequence of governmental attempts to ensure public well-being. Similarly, the set of values and life experiences of each individual colors his or her perception of what is or is not intrusive.

Evidently then, as for public involvement, evaluating pollution control with respect to nonintrusiveness is an endeavor fraught with difficulty. Nonetheless, given that nonintrusiveness is an important social value, we believe that an evaluation of the pollution control regulatory system would be incomplete without consideration of how that system measures up with respect to nonintrusiveness. At the very least, the criterion of nonintrusiveness warrants consideration because the degree to which a program is intrusive may contribute to that program's success or failure.

Definition of Intrusiveness

According to Webster's dictionary (Webster's 1988) intrusiveness is the state, quality, condition, or degree of being intrusive, where intrusive means tending to intrude, and to intrude means to thrust or force in or upon, especially without invitation, permission, or welcome. With the above definition in mind, we define "intrusiveness" to be that characteristic resulting in any unwelcome or uninvited imposition on the public, including direct cost, lost time, restricted options, invasion of privacy, or inconvenience.

Cost, considered elsewhere in this report, is an intrusive aspect of pollution control programs. Irrespective of other considerations, when a business sees its direct costs rise due to the requirements of a pollution control program, it is subject to an uninvited imposition. Similarly, an individual forced to pay a motor vehicle inspection fee experiences an uninvited direct cost.

Lost time, too, is often an uninvited imposition that impacts all sectors of the public. A large corporation experiencing a delay in bringing a product to market, an individual waiting in line at an automobile inspection center, and a small business owner who must fill out lengthy forms all might argue that the time loss imposed by government requirements is an imposition.

Pollution control programs may also lead to a restriction of the options available to the public, which is an uninvited imposition and thus intrusive. By restricted options, we mean limitations placed on the current or potential activities of a segment of the public. For example, certain Clean Water Act provisions limit property owners from developing areas that are "wetlands." Similarly, under the Toxic Substances Control Act, a large corporation may be prohibited from producing a new chemical or from marketing an existing chemical. Almost all pollution control programs place limitations on current activities, thereby restricting the options of the affected public.

Invasion of privacy is an imposition stemming from some pollution control programs. For example, communities that recycle sometimes hire inspectors to randomly dig through the trash of community members in order to enforce local recycling ordinances concerning the proper disposition of recyclables. Many affected individuals believe that such inspections amount to a substantial invasion of their privacy. A large corporation might also consider the requirements of a pollution control program a privacy violation. Companies concerned with the preservation of trade secrets might be reluctant to meet the reporting requirements of right-to-know legislation.

Finally, inconvenience is a relatively common intrusive attribute of pollution control programs. It is a notion meant to encompass intrusive aspects of pollution control programs that are not easily classified as lost time, direct cost, restricted options, or invasion of privacy. For example, the requirements of recycling programs often necessitate that households maintain several receptacles for the disposition of discarded materials such as plastic, glass, and metal. Though not a substantial problem for rural and suburban households, these recycling requirements often impose a major inconvenience on urban households, in that the latter typically live in smaller residences than their rural and suburban counterparts. The smaller residences poorly accommodate storage receptacles, which take up needed space and are an eyesore. Similarly, inconvenience may be imposed by recycling requirements on small food and beverage businesses that require them to accept and store recyclable bottles and cans.

Measuring Intrusiveness

There is no simple way to measure intrusiveness given its multiple forms, the numerous groups it affects, and its often subjective nature. An economist might attempt to size up the intrusiveness of a pollution control program by translating the various impositions of that program into the common denominator of dollars. Congress uses a political measure of intrusiveness, avoiding imposing regulatory requirements when a large number of people are affected. It relies on nonregulatory tools, such as education and voluntary programs, when dealing with problems whose solutions would require many people to act (such as radon in homes or water runoff from agricultural fields).

To provide some measure of intrusiveness, we developed a rudimentary "index of intrusiveness." For selected components of the pollution control regulatory system, we considered the degree to which each imposition type, as defined above, affects each of several segments of the public and state and local governments. In other words, for a given regulatory program, we asked the following question: "To what degree are citizens, small and medium businesses, large corporations, and state and local governments subject to direct costs, lost time, restricted options, privacy violation, and inconvenience?" Table 5 demonstrates the calculation of intrusiveness index values. The numerical values represent our best subjective estimates, but they have no basis in science or research.

Although we have presented a way to think about and rank the intrusiveness of pollution control programs, time and resource constraints circumscribed our effort. There are several obvious problems with our simple approach. First, we have segmented the public in an artificial manner, proposing categories that do not take into account the large array of affected groups. Second, by selecting a three-tiered strategy for rating the degree to which an intrusive attribute impacts a segment of the public, we have chosen a metric that lacks sensitivity. Third, we have neither presented nor developed guidelines for rating the impact of the various imposition types examined. Our cell values are best estimates based only on our knowledge of pollution control programs.. Fourth, we have equally weighted each imposition type for each affected public and we have given equal weight to imposition values across affected publics. Fifth, our selection of imposition types was limited to five in number; arguably, there exist other attributes of intrusiveness. (For example, there exists a literature on the "coerciveness" of regulatory programs. Arguably, we could have adopted coerciveness as a sixth attribute of intrusiveness.)

Despite these shortcomings, we can draw several preliminary conclusions concerning the intrusiveness of pollution control programs. First, the pollution control programs we examined fall along a wide spectrum in terms of their intrusiveness, with the National Environmental Policy Act being the least intrusive and the Clean Air Act the most. Second, the widely criticized Comprehensive Environmental Response, Compensation, and Liability Act is no more intrusive than most pollution control programs. Third, recycling programs, generally perceived very positively by the public, are surprisingly as intrusive as most other programs, indicating—not surprisingly—that criteria other than nonintrusiveness influence the public's perception of pollution control programs.

Table 5. Intrusiveness Index Calculations.

	RCRA	CERCLA	CWA	CAA	SDWA	TSCA	FIFRA	NEPA	Recycl-ing
Citizens									
Lost time	1	1	1	2	1	1	1	1	2
Direct cost	1	1	1	3	2	1	1	1	1
Restricted options	1	1	2	1	1	1	1	1	1
Privacy violation	1	1	1	1	1	1	1	1	1
Inconvenience	1	1	1	2	1	1	1	1	3
Small and medium businesses									
Lost time	1	1	1	2	1	2	1	1	2
Direct cost	2	1	1	2	1	1	1	1	1
Restricted options	1	1	1	2	1	1	1	1	1
Privacy violation	1	1	1	1	1	1	1	1	1
Inconvenience	1	1	2	2	1	2	1	1	2
Large companies									
Lost time	1	1	1	2	1	2	1	1	1
Direct cost	2	2	2	3	1	1	2	1	1
Restricted options	1	2	2	2	1	1	2	1	1
Privacy violation	1	1	1	1	1	1	1	1	1
Inconvenience	1	1	2	2	1	2	2	1	1
State and local governments									
Lost time	1	1	2	3	2	1	1	1	1
Direct cost	1	1	2	2	3	1	1	1	2
Restricted options	2	2	2	2	2	1	2	1	2
Privacy violation	1	1	1	1	1	1	1	1	1
Inconvenience	1	1	2	2	2	1	2	1	1
Total raw index value	23	23	29	38	26	24	25	20	27

Notes: Ratings: Most Intrusive: 3; Medium Intrusiveness: 2; Least Intrusive: 1. The ratings are the subjective opinions of the authors.

A rating of 3 indicates that a particular intrusiveness attribute is a *very substantial* imposition on a given public, 2 is indicative of a *substantial* imposition, and 1 indicates a *minor* imposition or *none* whatsoever.

For a given pollution control program, an index value is arrived at by summing across publics the cell values associated with each imposition type. In each column, twenty values ranging from one to three are added together to develop a raw intrusiveness score. When repeated for all of the selected programs, a full set of index values is established. The developed index values can be compared to establish an intrusiveness ranking of the programs.

Little recognition has been given to intrusiveness by analysts of pollution control programs, despite solid evidence indicating that nonintrusiveness and other social values influence program success. Increased understanding and awareness of nonintrusiveness as a social value, a more refined definition of intrusiveness, and improved measures of it might together provide policymakers with important information.

The Paperwork Burden

The Paperwork Reduction Act requires federal agencies to provide the Office of Management and Budget with estimates of the time necessary for businesses and state and local governments to complete required reports. Such data assist OMB in assessing federal agencies' reporting requirements, which must receive the OMB stamp of approval prior to being imposed on businesses and governmental entities. Data collected from all federal agencies are aggregated by OMB and incorporated into a document entitled the "Information Collection Budget," which is annually submitted to Congress.

Many of the entities subject to the paperwork burden of reporting requirements consider the requirements to be an intrusion that manifest some if not all of the various attributes of intrusiveness outlined above. In most cases, the nonpaperwork requirements of a program's regulations are actually more intrusive—that is, both the number and the degree of associated impositions are greater. However, paperwork burdens have, for a variety of reasons, often become a symbol of the intrusiveness of government programs and thus may be perceived by regulatees to be more offensive than nonpaperwork burdens. Perhaps most offensive to regulatees is the fact that reporting requirements do not directly contribute to the prevention or reduction of pollution. Also, unnecessary duplication is often obvious in the reporting requirements imposed by various government offices.

Using 1994 data obtained from the Chemical Manufacturers Association (CMA), we compared the reporting requirements of pollution control programs. The CMA data were generated from raw data on information collection budgets submitted by EPA to the Office of Management and Budget. (OMB does not include paperwork burden data broken down by statutory program in its Information Collection Budget. Due to time constraints, we used CMA data rather than aggregate raw data ourselves.) Next, using data obtained from recent OMB information collection budgets, we considered the reporting requirements associated with all EPA-administered programs and compared them with those of other federal agencies, examining how reporting requirements have evolved over several years.

Table 6. Annual Reporting Requirements of Selected EPA-Administered Statutes.

EPA statute	Annual reporting hours (thousands)	Annual responses (thousands)	Number of report types
Clean Water Act	29,812	1,249	7
Clean Air Act	9,331	839	40
Resources Conservation and Recovery Act	7,174	546	16
Superfund Amendments and Reauthorization Act	4,334	81	3
Federal Insecticide, Fungicide, and Rodenticide Act	3,195	204	10
Toxic Substances Control Act	512	22	12
Comprehensive Environmental Response, Compensation, and Liability Act	280	37	4
Safe Drinking Water Act	11	4	1

Source: CMA 1996.

Table 6 presents data concerning the estimated 1994 reporting requirements of eight EPA-administered statutes, ranked according to total number of annual reporting hours (CMA 1996, 5–14). Based on the total number of annual reporting hours required by business and state and local governments, the three most intrusive sets of reporting requirements appear associated with the Clean Water Act, the Clean Air Act, and the Resources Conservation and Recovery Act. Interestingly, the regulatory programs associated with the Clean Water Act and the Clean Air Act were ranked by our intrusiveness index as the two most intrusive pollution control programs.

Table 7 presents data concerning the estimated reporting requirements of ten federal agencies and departments over an eight-year period. In 1995, the selected agencies had the largest estimated paperwork burdens. (Our calculations are based on estimated total information collection burdens for FY1989–1996 as reported by the Office of Information and Regulatory Affairs of the Office of Management and Budget.)

Table 7 shows some interesting results. First, as measured by the total number of annual reporting hours, the 1995 reporting requirements associated with EPA regulations rank eighth among the selected federal entities, with the Treasury Department, the Department of Labor, the Department of Defense, the Securities and Exchange Commission, the Department of Health and Human Services, the Federal Trade Commission, and the Department of Agriculture all imposing higher paperwork burdens on regulatees. EPA's relative position in previous years was typi-

Table 7. Annual Reporting Requirements of Selected Agencies (thousands of hours).

Agency	1988	1989	1990	1991	1992	1993	1994	1995[a]	% change 1988 to 1995
Treasury	4,162,059	4,270,402	5,162,768	5,572,260	5,781,073	5,075,640	5,079,831	5,331,298	28.09%
DOL	129,165	109,270	73,871	61,537	53,291	47,552	48,429	266,448	106.28%
DOD	263,169	264,018	226,273	228,614	228,994	231,041	217,053	205,848	-21.78%
SEC	63,306	57,855	62,292	63,859	63,771	66,142	201,631	191,527	202.54%
HHS	69,208	158,430	242,996	248,236	271,685	289,011	351,497	152,616	120.52%
FTC	65,564	64,910	64,908	7,597	2,510	2,531	5,157	146,149	122.91%
USDA	70,233	76,898	77,769	82,129	86,665	90,345	93,971	131,001	86.52%
EPA	69,174	77,456	82,459	58,486	61,104	60,955	69,253	103,066	49.00%
DOT	87,858	109,680	109,935	111,774	77,996	79,003	95,414	91,023	3.60%
DOE	36,840	38,575	40,823	44,200	49,863	52,806	65,337	57,555	56.23%
Totals for all agencies	5,155,007	5,351,826	6,292,854	6,642,844	6,843,109	6,190,689	6,411,254	6,900,932	33.87%

[a] In FY1995, third-party and public disclosures were covered by the paperwork burden accounting system.

Notes: DOL: Department of Labor; DOD: Department of Defense; SEC: Securities Exchange Commission; HHS: Health and Human Services; FTC: Federal Trade Commission; USDA: United States Department of Agriculture; EPA: Environmental Protection Agency; DOT: Department of Transportation; DOE: Department of Education.

The different totals for 1994 EPA annual reporting hours in Tables 6 and 7 can be explained by the fact that Table 6 only covers eight EPA statutes, while Table 7 covers all EPA statutes.

Source: U.S. OMB 1996.

cally seventh or eighth, and never higher than fifth. Second, during the eight-year period, EPA's paperwork burden grew by about 50 percent. The paperwork burdens of six other agencies grew at an even higher rate, some more than doubling in size. (Some of the increase over this period in EPA's paperwork burden, and the increases seen in other agencies' paperwork burdens, can be explained by the inclusion in FY1995 of third-party and public disclosures in the paperwork burden accounting system.)

Many of the selected agencies are larger than EPA in terms of budget, employees, and other measures, but given recent, heated denunciations of EPA, the above figures provide some markers for understanding the paperwork burden imposed by EPA. Although the EPA-imposed paperwork burden is not small, for an agency with cabinet-level responsibilities that burden is not as egregious as some critics have claimed. Though substantial, it is not unusually intrusive when compared with the paperwork burdens imposed by other major federal agencies and departments.

ENVIRONMENTAL JUSTICE

The environmental justice movement can be viewed as a form of political mobilization in which affected communities challenge local industries and the public agencies that regulate them to reduce or eliminate pollution and other environmental hazards. In contrast with the more libertarian view that environmental regulations are intrusive and best kept to a minimum, advocates for environmental justice have demanded increased regulatory oversight and more stringent environmental protection measures to better protect minority and low-income communities and workers. (For a review of the general issue, see Ringquist in Vig and Kraft 1997.)

To understand the challenges environmental justice poses to the pollution control system and to evaluate the response of the system to these demands, we need to be cognizant of the currents that have informed the movement. As early as the mid-1970s, local labor unions and community activists in cities across the country had begun forming committees on occupational safety and health to document worker exposure to hazardous chemicals in the workplace. This effort was prompted in part by federal initiatives, as the creation of the Occupational Safety and Health Administration in 1970 along with the National Institute of Occupational Safety and Health facilitated research on workplace exposures and provided new avenues for citizen participation. In 1978, with the revelations at Love Canal of a residential community mired in toxic wastes, advocates for the committees began to establish strong ties with community groups across the country, calling for right-to-know laws and other measures. By the early 1980s, an estimated ten thousand local groups, many from poor and minority communities, had become part of a nationwide anti-toxics movement, coordinated by the Citizens' Clearinghouse on Hazardous Waste and the National Toxic Campaign (see Gottlieb 1995; Szasz 1994).

The burgeoning environmental justice movement was not confined to the cities. Under the banner of civil rights and within the tradition of civil disobedience, rural African-American communities had begun to protest the siting of hazardous facilities in their communities. In 1982, the General Accounting Office studied hazardous waste landfills in eight southeastern states at the request of Representative Walter Fauntroy, who was arrested for demonstrating in opposition to the siting of a polychlorinated biphenyl disposal landfill in a predominantly black and poor rural county in North Carolina. The study found that African-Americans comprised the majority of the population at three of the four off-site hazardous waste sites in the region, but could not say if the findings reflected national patterns or demonstrated racist intent in siting procedures (U.S. GAO 1983). (More recently, Vicki Been (1995) has shown that when one looks at the demographics surrounding the sites at the time of the siting, three out of four communities at the sites were not predominantly African-American. She suggests that due to market forces, African-Americans were drawn to lower-cost housing near the sites in the intervening period.)

To extend the scope of the GAO report to the national level, the United Church of Christ's Commission for Racial Justice analyzed the demographics of communities where toxic waste sites and hazardous waste treatment and disposal facilities were located. The release of the report, *Toxic Wastes and Race in the United States* (UCC 1987), brought the issue of environmental racism to national attention. The study found that the mean percentage of racial minorities and low-income groups in communities with one or more operating hazardous waste facilities was twice that of communities with no facilities (24 percent versus 12 percent) (UCC 1987, 63). The study also found that race was a stronger explanatory variable than class, a point taken up by the Michigan Coalition, a small group of academicians and civil rights leaders formed following the University of Michigan's "Conference on Race and the Incidence of Environmental Hazards." In response to the lobbying efforts of the Michigan Coalition, then-EPA Administrator William Reilly formed the Environmental Equity Working Group in the summer of 1990 to evaluate the evidence that racial minorities and low-income groups bore a disproportionate burden of risk and to review EPA's performance in addressing environmental inequities.

Framing the Problem

Two years later, the EPA working group's report, *Environmental Equity: Reducing Risk for All Communities*, concluded that racial minorities and the

poor suffer higher than average potential exposures to selected air pollutants, hazardous waste facilities, and pesticides in the workplace (U.S. EPA
1992b, 3). The report also noted that while clear differences exist between
racial groups in terms of disease and death rates, environmental and
health data are not routinely collected and analyzed by income or race. In
the absence of such data, the report was careful to point out, it is not possible to say whether such groups are exposed to higher levels of pollution.
To address these problems, the report recommended that EPA should
give greater attention to issues of environmental equity, revising risk
assessment procedures to improve the "characterization of risk across
population, communities, or geographic area" (U.S. EPA 1992b, 4).

The report was sharply rebuked by many in the environmental justice movement. The agency view of environmental equity, it was argued,
was confined to risk analysis, relied on expert opinion to define and
assess risks rather than the experience and needs of affected communities, and sought to analyze the distribution of environmental risks across
population groups with little regard for the institutional and structural
forces that have contributed to the subordination of minority and low-
income communities (see U.S. EPA 1992b, Supporting Document). While
EPA claimed that "a person's activity pattern is the single most important
determinant of environmental exposures for most pollutants" (U.S. EPA
1992b, 7), representatives of the minority community argued that EPA, in
a display of circular reasoning, was essentially attributing exposure to the
very factors that lead to the disparities for minorities and the poor in the
first place: limited mobility, reduced housing options, fewer employment
opportunities, diet, and poorer municipal services. The working group
report, it was argued, had missed the point, that unfair outcomes
occurred because of institutionalized forms of racism. Decades of discrimination, according to a noted critic, have left "blacks and lower-income
groups trapped in polluted environments," and the agency itself, it was
alleged, bore responsibility for perpetuating environmental injustice by
defining its mission narrowly (seeking risk reduction rather than promoting social welfare). Critics of the report called on EPA to take targeted and
aggressive actions to eliminate—not simply redistribute—risks from
industrial pollution and to devise procedures for siting noxious facilities
that could lead to fairer outcomes.

From Equity to Justice

Less than seven months after the working group's report was released, a
new administration was in the White House. EPA's Office of Environmental Equity, established by Reilly to provide interagency coordination of
equity programs, was renamed the Office of Environmental Justice. This

was more than just a change in office stationery. Shortly after taking office, EPA Administrator Carol Browner listed environmental justice as one of the top priorities for her term at EPA and staked out a broader view of the agency's responsibilities in the area of environmental justice, stating that "incorporating environmental justice into everyday activities and decisions will be a major undertaking. Fundamental reform will be needed in agency operations" (U.S. EPA 1994). To help put this view into practice, EPA established the National Environmental Justice Advisory Council, which includes representatives from community groups; universities; environmental organizations; industry; and state, local, and tribal governments. A number of the council's members, such as Robert Bullard and Charles Lee, were among the sternest critics of the working group report. Under the new administration, the council was asked to help devise an environmental justice strategy for the agency, one that ultimately could influence other federal agencies and departments.

On February 11, 1994, President Clinton issued Executive Order 12898, Federal Actions to Address Environmental Justice in Minority Populations and Low-Income Populations. The order required all federal agencies conducting activities that substantially affect human health and the environment to make environmental justice a priority and directed them to develop "environmental justice strategies" within a year that identified and addressed adverse human health and environmental effects of their programs, policies, and activities on minority and low-income communities.

EPA published its Environmental Justice Strategy in April 1995. The two goals specified in the document are an odd mixture of activism and regulatory caution. The first goal requires the agency to examine its internal administrative procedures for evidence that its programs and policies contribute to environmental injustices: "No segment of the population, regardless of race, color, national origin, or income [should], as a result of EPA's policies, programs and activities, suffer disproportionately from adverse human health or environmental effects, and all people [should] live in clean, healthy and sustainable communities" (U.S. EPA 1995a, 1).

Clearly, this language suggests the activist orientation that the environmental justice movement had called for, but it stops short of meeting a central demand of the movement, namely that EPA recognize its responsibility to intervene in matters that contribute to unfair environmental outcomes such as local zoning, redlining, and unfair housing practices—activities that traditionally have been immune to EPA regulation and outside the agency's mission.

The second goal commits the agency to increase public participation in environmental decisionmaking: "Those who live with environmental decisions—community residents, State, Tribal, and local governments, environmental groups, businesses—must have every opportunity for public participation in the making of those decisions. An informed and involved community is a necessary and integral part of the process to protect the environment" (U.S. EPA 1995a, 102). The critical elements of strong public participation in the agency's Environmental Justice Strategy are access to extensive and useful information, increased technical assistance grants to low-income and minority communities, early discussions with affected communities, and partnerships with academic institutions (including historically black colleges and universities, hispanic serving institutions, and tribal colleges). It is unclear, however, if the goal of increased opportunities for the public to participate in environmental decisions will lead to increased opportunities for the public to determine final policy outcomes, a key demand of many environmental justice activists (Ferris 1994; Miller 1994; Shepard 1994).

Since the release of the strategy, EPA has established numerous environmental pilot projects, both at headquarters and in the regions, that range from research in risk communication strategies to explain pollution risks to urban minority communities, to projects that involve linking cleanups at hazardous waste sites to minority hiring requirements. In addition, EPA program offices have developed training programs for both EPA and state regulatory officials that examine how environmental justice concerns can be integrated into routine agency functions. The agency established a small grants program in 1994 that distributed $3 million to some 170 community-based/grassroots organizations and tribal governments to assist them in public outreach and to enable them to employ technical experts to analyze and interpret environmental data. The following year, building on this momentum, the agency developed the Community/University Partnership program to expand training and educational opportunities for local communities and to develop models of collaboration between universities and affected communities.

Environmental Injustice: The Evidence

As these initiatives indicate, environmental justice has become a top priority for EPA under the current administration, but the results of empirical research to characterize environmental injustice and to prove environmental discrimination have been mixed. During the last decade, many studies have attempted to demonstrate the existence of inequities in the social distribution of toxic facilities and unwanted local land uses. These studies, conducted at both national and regional levels, employed readily available spatial units of analysis such as zip codes, counties, and census tracts, and used census data from a specific year to compare the socioeconomic status of communities hosting unwanted facilities with those that did not (Bullard 1993; Burke 1993; Goldman 1991; UCC 1987; U.S. GAO 1983).

At the national level, the United Church of Christ's report *Toxic Wastes and Race in the United States* (UCC 1987) and an update of the report, *Toxic Wastes and Race Revisited* (UCC 1994), focused on commercial hazardous waste facilities and uncontrolled waste sites as environmental threats. Using zip codes as the areal unit, the study found that "race proved to be the most significant among variables tested in association with the location of hazardous waste facilities" (UCC 1987, xiii). The results of these studies were challenged by an analysis conducted by the Social and Demographic Research Institute at the University of Massachusetts (Anderton and others 1994). These investigators studied the same facilities examined in the two United Church of Christ reports but employed census tracts within a national sample of metropolitan areas rather than zip codes. They found no statistically significant differences between the mean percentages of African-Americans or Hispanics in host and nonhost tracts. Their study was criticized for limiting its comparison to nonhost tracts in metropolitan areas or rural counties with one commercial hazardous waste facility (UCC 1994, 3). Critics argued that without an accurate model of how these facilities are sited, all tracts should be considered as possible alternatives for the site, not just those in the same metropolitan area.

To explain the discrepancies between the University of Massachusetts and United Church of Christ analyses, Been (1995) compared demographic variables between all commercial hazardous waste host tracts and all populated nonhost tracts within the continental United States. Moreover, instead of comparing population means, as both earlier studies had done, her study examined how distribution of facilities is related to the distribution of the population around the mean in host tracts. Been's findings differed from the findings of the earlier reports, showing that while many neighborhoods with percentages of African-Americans greater than the national average bear a disproportionate share of the nation's facilities, almost as many lower-income neighborhoods with few African-Americans are also bearing a disproportionate share. Environmental injustice, the report concluded, is a "complicated entanglement of class, race, educational attainment, occupational patterns, relationships between the metropolitan areas and rural or non-metropolitan cities, and possibly market dynamics" (Been 1995).

The ambiguity of results found at the national level also can be found in state- and metropolitan-level studies. In a statewide analysis of toxic air releases and hazardous waste generation in forty-six counties in South Carolina, Cutter (1994, 41) found that the most affected residents, based on proximity to the sources of the releases, lived in racially mixed communities with average incomes. A more detailed study conducted at the block level of commercial hazardous waste facilities in South Carolina found that such facilities are clustered around urbanized areas with high population densities but are not disproportionately located in minority or economically disadvantaged communities (see Holm 1994). Studies at the metropolitan level suggest an equally complicated picture. Burke's analysis of Toxic Release Inventory sites in Los Angeles shows a clear correlation between minority percentage and the number of sites within census tracts (Burke 1993). Similarly, Glickman and Hersh found that Toxic Release Inventory sites in the greater Pittsburgh area are disproportionately located in African-American neighborhoods, but commercial hazardous waste facilities and facilities that store extremely hazardous substances having the potential to form a toxic vapor cloud when accidentally released are more likely to be found on the outskirts of the city along major transportation arteries, located in white working-class neighborhoods (Glickman and Hersh 1995).

What constitutes environmental injustice, then, is difficult to answer. The spatial unit of analysis chosen and the implicit definition of an affected community can clearly have profound impacts on the results of an environmental justice analysis. For example, to define a community at the zip code level, as the United Church of Christ studies have done, can mask finer patterns of spatial and social segregation nearer to the site. Nor can an affected community be defined simply as those individuals living near a hazard. Other possible definitions of an affected community could be related not to proximity to a hazard but to risk, with the boundaries of an affected community coincident to a concentration contour of an air pollutant or to the users of groundwater from a contaminated aquifer. An affected community could be even more dispersed spatially if one chose to define such a community by the degree of concern about a potential hazard or risk. The social boundaries of this concern may far exceed the area immediately adjacent to the hazard and the definition of the affected community may change over time.

Another difficulty in defining environmental justice is that much of the research thus far has been limited to examining the proximity of minority and low-income communities to hazardous waste facilities or Toxic Release Inventory facilities, with proximity serving as a substitute for risk. A number of potentially more threatening environmental risks to low-income and minority communities (contaminated drinking water, radon, indoor air pollution) have not been studied. Without such information, EPA has an incomplete picture of the risks that communities face, the spatial and social distribution of these risks, and the risk or exposure thresholds that constitute environmental discrimination.

Further muddying the waters is that environmental injustice is seen by many environmental justice advocates not only as inequitable envi-

184 POLLUTION CONTROL IN THE UNITED STATES

ronmental outcomes along class and racial lines but as a process in which poor and minority communities get more of the costs than benefits of environmental policies because of their lack of political power, the differential enforcement of environmental regulations, and basic social inequalities. Yet few studies have fully explained the factors that influence the siting of commercial hazardous waste and similar facilities, or traced the root causes of environmental discrimination in an historical context, or assessed whether government agencies have played a role in perpetuating or creating environmental injustice through varying enforcement of regulations. The one oft-cited exception is a 1992 study in the *National Law Journal,* which found that EPA set stiffer financial penalties for violations at Superfund sites from 1985 through 1991 that occurred in predominately white communities than in minority communities (Lavelle and Coyle 1992).

Evaluating EPA's Efforts

Little systematic evaluation has been conducted to measure how well the programs and initiatives of EPA's Environmental Justice Strategy have worked to improve environmental quality for minority and low income communities. This is due in part to the difficulty of knowing what to measure and what baseline data to use to assess the various initiatives. What hazards represent the most serious health threat to the general public and to minority and poor communities? If risk is discarded as a common metric to compare various environmental problems, on what basis should the agency set environmental priorities? In the place of explicit procedures for establishing tradeoffs between the costs of environmental protection and other societal goals, will environmental priorities be set by groups that can better mobilize resources to the detriment of minority communities?

While environmental justice advocates have not pushed for a common definition of "environmental justice" in the interest of building a movement, EPA is in the unfortunate position of trying to fashion an environmental justice strategy without a clear, operational definition of environmental justice. This presents the agency with a number of difficulties.

The first difficulty is tied to redefining the mission of the agency. To what extent can the pollution control regulatory system, with EPA at its hub, use environmental regulations to promote social equity? The obstacle here is related to rival conceptions of justice that inform, on the one hand, the environmental justice movement and, on the other, the work of the pollution control regulatory system. For the last twenty-five years, the overarching strategy of environmental protection in this country has been based not on a standard of justice that assumes government regulation must be directed to improving the conditions of the most destitute members of society, but on a utilitarian principle that considers the level of aggregate benefits across society (that is, the greatest happiness for the greatest number of people).

The utilitarian principle is incorporated into environmental policies through such tools as benefit-cost analysis and, more recently, comparative risk. The aim of the system in principle is to reduce risk and to protect the natural environment through strategies that move toward allocative efficiency. In benefit-cost analysis, environmental policies should maximize net benefits to society as a whole rather than promote the rights of particular groups. Comparative risk, like benefit-cost analysis, is premised on the notion that environmental protection involves trade-offs and that, with limited resources available, EPA and state regulatory agencies should give the highest priority to the most serious risks to human health and the environment. In rankings of risk conducted by scientific panels (see Chapter 6), those hazards that often appear at the top of the environmental justice agenda—toxic dumps, commercial waste handlers—have been given less importance. It is unclear how EPA or state and local agencies can integrate into their regulatory programs divergent perceptions about the severity of environmental problems.

Second, while environmental justice advocates have called on EPA to take into account a fuller range of environmental hazards to which minority residents are often exposed and to devise methods to model cumulative and synergistic risks (Bullard 1993; Mohai and Bunyan 1992), the framework of federal pollution control laws is so fragmented that this demand is unlikely to be met. The single-medium focus has tended to shift pollution from one medium to another, sometimes exacerbating the situation or failing to detect the problem altogether. The current regulatory framework, as noted in Chapter 2, does not readily allow EPA to undertake the sort of integrated risk assessment and environmental management that the environmental justice movement and others have called for.

Third, equal enforcement of environmental regulations is a particular concern for the environmental justice movement after a series of articles appeared in the *National Law Journal* in 1992 that compiled substantial evidence of less-stringent regulatory oversight and enforcement of the Comprehensive Environmental Response, Compensation, and Liability Act and the Resource Conservation and Recovery Act in minority communities (Lavelle and Coyle 1992). Yet the authority to monitor compliance and to initiate enforcement actions against firms or other entities operating out of compliance has been relegated by EPA to the states or to local regulatory agencies under a number of statutes (see Chapter 4). Similarly, the siting of waste facilities is typically under the control of state or local land-use law, and the states and localities have historically been hostile to attempts by the federal government to intervene in land-use decisions.

Typically, EPA's role in permitting comes after the site has been selected, and it primarily involves technical considerations.

For all these reasons, EPA is unlikely to have a major impact on the concerns of the environmental justice movement. In any struggle over the equitable distribution of resources in American society, environmental resources are not likely to be front-and-center.

OTHER SOCIAL VALUES

A large number of other social values are in some way relevant to pollution control. For example, the literature on risk perception contains a number of social values (such as involuntary risks being considered less acceptable than voluntary risks) that influence people's risk perceptions (see Slovic in Krimsky and Golding 1992). We do not have enough space or time in this report to deal with all relevant values.

One broad value that should be mentioned is cooperation. American society has long been ambivalent about the value of cooperation—our society values competition and conflict as much as or more than cooperation. Our economic system puts a high value on competition, and our legal system is based on conflict as a method for arriving at decisions.

On net, we would say that the pollution control regulatory system has fostered conflict more than cooperation, although it has contributed to both. EPA was one of the pioneers in using regulatory negotiation, and several state-level efforts have demonstrated new avenues for cooperative problem solving (see, for example, Snow 1996). However, environmental law has contributed greatly to the litigious nature of American society, and much pollution control relies on citizen suits, liability, and other conflict-based mechanisms to provide the incentives for preventing or controlling pollution. EPA has initiated several voluntary programs, such as the 33/50 Program, to reduce toxic emissions, but these programs have been peripheral to the main pollution control activities (Davies and Mazurek 1996).

Many of the value questions related to regulation are connected to the underlying question of trust. Whether the federal government, EPA, or state governments deserve the trust of the American people is a question this report will not attempt to answer. Without a minimal degree of trust, however, governance of any kind, including governance to control pollution, is impossible. The continuing decline of trust in government and all other institutions may be the single most important problem now facing the country. The extent of this distrust is just one of many ways in which the United States differs from other industrialized countries.

The pollution control system also must be given at least partial credit for establishing and encouraging a broad ethical belief in a clean environment. The underlying views that have provided the bedrock support for the system were created, at least in part, by the system itself. State and federal officials provided much of the information and led many of the education efforts that first brought environmental quality to public attention. The continuing efforts of environmental organizations, businesses, and government agencies have resulted in environmental protections being not a transitory fad but instead a permanent and important value held by most of the American people.

CONCLUSIONS

Public Involvement

The pollution control regulatory system enables the public to influence outcomes at many points. We have discussed notice-and-comment rule-making, environmental permitting, ensuring government and industry accountability, environmental litigation, recycling, and information exchange. Although specific forms of public involvement may be flawed in some manner, the diverse array of public involvement mechanisms associated with the pollution control regulatory system is impressive.

Public involvement in the notice-and-comment process and in the environmental permitting process would benefit from the increased use of mechanisms that promote early public involvement. Our analysis found that EPA has not used some of these mechanisms as frequently as other federal agencies. We also found that EPA's use of one important early public involvement mechanism—regulatory negotiation—has declined in recent years. And while there are no formal mechanisms in place to engage the public early in the permitting process, recent EPA reports have indicated the necessity of doing so (U.S. EPA 1995b, 1996).

Increased use of advisory committees would promote dialogue between EPA and the public, which would enhance public involvement in both rulemaking and permitting. Moreover, the absence of such dialogue is an impediment to cultivating a more sophisticated public understanding of environmental decisionmaking. The requirement that agencies obtain permission from the Office of Management and Budget to establish new advisory committees, along with the negative White House perception of such committees, serves as a deterrent to the formation of advisory committees. Since OMB permission to form new advisory committees is required under the Federal Advisory Committee Act, Congress would likely have to pass amending legislation to remove or limit OMB's authority before advisory committees could increase in number. Such an action would be appropriate and would enhance public involvement.

Some of the most important gains in environmental protection have come through the public's use of environmental litigation. As we have described, this form of public involvement has been instrumental to the maturation of other forms of public involvement in the pollution control regulatory system. In recent years, however, some observers have raised concern over whether public access to the courts will be diminished, due to the adoption, in their view, of a more restrictive interpretation by the courts of "standing" to sue. Whether a more restrictive interpretation of standing represents an impediment to citizen-initiated environmental litigation or just an inconvenience is unclear at this time.

Nonintrusiveness

The social value of nonintrusiveness is very important but hard to measure and analyze. We have compared major components of the pollution control regulatory system and arrived at preliminary conclusions concerning their relative intrusiveness, compared and ranked reporting requirements in terms of intrusiveness, and assessed the intrusiveness of EPA reporting requirements vis-à-vis those of other federal agencies. We have established a working definition of intrusiveness and developed a rudimentary "index of intrusiveness" that could be refined and expanded.

Based on preliminary results, we judge the Clean Air Act and Clean Water Act to be the most intrusive components of the pollution control regulatory system, perhaps not unexpectedly. We find the reporting requirements of the same two acts to be the most intrusive reporting requirements of EPA-administered statutes. However, despite the claims of critics, EPA's burden is neither overly nor unusually intrusive among the paperwork burdens imposed by federal agencies.

Environmental Justice

The environmental justice movement has been able to force EPA and other federal agencies to consider the impact of their environmental policies and administrative practices on minority and low-income communities. The movement has forged alliances with civil rights activists, lawyers, academicians, and scientists. In the process, its demand for equal environmental protection for minority and low-income communities has become a broader demand for pollution prevention, community participation—some would say control—in the design and control of environmental programs, and a rejection of environmental priority setting based solely on science and technical risk assessment. EPA has been most successful incorporating environmental justice into its policies and activities by building the capacity of local communities through partnerships with local universities and awarding grants for technical assistance and outreach. The agency also has put considerable effort into educating its staff about environmental justice, and has set up administrative structures at headquarters and in the regions to direct attention to environmental justice issues. However, the utilitarian basis of the statutes and the regulatory framework under which the agency operates run counter to EPA's stated intentions to make environmental justice a high priority.

REFERENCES

Administrative Conference of the United States. 1995. *Negotiated Rulemaking Sourcebook.* Washington, D.C.: U.S. GPO.

Anderton, Douglas, and others. 1994. Environmental Equity: Evaluating TSDF Siting Over the Past Two Decades. *Waste Age* (July): 83–100.

Barbour, Ian G. 1980. *Technology, Environment, and Human Values.* New York: Praeger.

Been, Vicki. 1995. Analyzing Evidence of Environmental Justice. *Journal of Land Use and Environmental Law* 11(1): 21.

Bell, Ruth. 1992. Environmental Law Drafting in Central and Eastern Europe. *Environmental Law Reporter* 22(9, September): 10601.

BNA (Bureau of National Affairs). 1988. *U.S. Environmental Laws.* Washington, D.C.: BNA.

———. 1996. IG Report Says Agency Should Reduce Amount It Spends on Advisory Committees. *Environment Reporter,* May 10, 1996, p. 230.

Bobertz, Bradley. 1991. *Public Participation in Environmental Regulation.* Washington, D.C.: Environmental Law Institute.

Bowman, Margaret, ed. 1992. *The Role of the Citizen in Environmental Enforcement.* Washington, D.C.: Environmental Law Institute.

Bullard, Robert D. 1993. Anatomy of Environmental Racism and the Environmental Justice Movement. In Robert Bullard (ed.), *Confronting Environmental Racism: Voices from the Grass Roots.* Boston: South End Press.

Burke, Lauretta. 1993. *Environmental Equity in Los Angeles.* Technical Report 93-6. National Center for Geographic Information and Analysis (NCGIA). Santa Barbara, California: University of California.

Cardozo, Michael H. 1981. The Federal Advisory Committee Act in Operation. *Administrative Law Review* 33(1): 1–62.

CMA (Chemical Manufacturers Association). 1996. *Compliance Reporting Costs for EPA-Administered Environmental Laws: 1994.* Federal Relations Department, Economics Division. Washington, D.C.: CMA.

Cutter, S. L. 1994. The Burdens of Toxic Risks: Are They Fair? *Business and Economic Review* (41): 3–7.

Davies, J. Clarence, and Jan Mazurek. 1996. *Industry Incentives for Environmental Improvement: Evaluation of U.S. Federal Initiatives.* Washington, D.C.: Global Environmental Management Initiative.

Elliott, E. Donald. 1992. Re-Inventing Rulemaking. *Duke Law Journal* 41: 1490, 1492–93.

Ferris, Deeohn. 1994. Communities of Color and Hazardous Waste Cleanup: Expanding Public Participation in the Federal Superfund Program. *Fordham Urban Law Journal* 21 (3, Spring): 671–87.

Glickman, T. S., and R. Hersh. 1995. *Evaluating Environmental Equity: The Impacts of Industrial Hazards on Selected Social Groups in Allegheny County, Pennsylvania.* Discussion Paper 95-13. Washington, D.C.: Resources for the Future.

Goldman, Benjamin A. 1991. *The Truth About Where You Live: An Atlas for Action on Toxins and Mortality.* New York: Random House.

Gottlieb, Robert, ed. 1995. *Reducing Toxics: A New Approach to Policy and Industrial Decisionmaking.* Washington, D.C.: Island Press.

Holm, D. M. 1994. *Environmental Inequities in South Carolina: The Distribution of Hazardous Waste Facilities.* Master's thesis. Department of Geography, University of South Carolina.

Kraft, Michael. 1996. *Environmental Policy and Politics: Towards the 21st Century.* New York: HarperCollins College Publishers.

Krimsky, Sheldon, and Dominic Golding, eds. 1992. *Social Theories of Risk.* Westport, Connecticut: Praeger.

Landy, Marc K., Marc J. Roberts, and Stephen R. Thomas. 1994. *The Environmental Protection Agency: Asking the Wrong Questions, From Nixon to Reagan.* New York: Oxford University Press.

Lavelle, M., and M. Coyle. 1992. Unequal Protection: The Racial Divide in Environmental Law. *National Law Journal,* September 21, Supplement.

McElfish, James M. 1993. Drafting Standing Affidavits After *Defenders*: In the Court's Own Words. *Environmental Law Reporter* 23(1, January): 10026–30.

Miller, Vernice. 1994. Planning, Power, and Politics: A Case Study of the Land Use and Siting History of the North River Water Pollution Control Plant. *Fordham Urban Law Journal* 21(3, Spring): 707–22.

Mohai, Paul, and Bryant Bunyan. 1992. Environmental Racism: Reviewing the Evidence. In Bryant Bunyan and Paul Mohai (eds.), *Race and the Incidence of Environmental Hazards: A Time for Discourse.* Boulder, Colorado: Westview Press.

Shepard, Peggy. 1994. Issues of Community Empowerment. *Fordham Urban Law Journal* 21(3, Spring): 739–55.

Snow, Donald. 1996. River Story. *Chronicle of Community* 1(1): 16–25.

Szasz, Andrew. 1994. *Ecopopulism: Toxic Waste and the Movement for Environmental Justice.* Minneapolis: University of Minnesota Press.

UCC (United Church of Christ). 1987. *Toxic Wastes and Race in the United States.* Commission for Racial Justice. New York: UCC.

———. 1994. *Toxic Wastes and Race Revisited.* Commission for Racial Justice. New York: UCC.

U.S. CEQ (Council on Environmental Quality). 1993. *Environmental Quality: Twenty-Fourth Annual Report.* Washington, D.C.: U.S. GPO.

———. 1997. *Environmental Quality: 25th Anniversary Report.* Washington, D.C.: U.S. GPO.

U.S. DOE (Department of Energy). 1996. *Strengthening the Department of Energy's Commitment to Openness through the Freedom of Information Act.* Openness Press Conference Fact Sheets. Washington, D.C.: U.S. DOE.

U.S. EPA (Environmental Protection Agency). 1992a. *Environmental Equity: Reducing Risk for All Communities.* Volume I. Office of Policy, Planning, and Evaluation. Washington, D.C.: U.S. EPA.

———. 1992b. *Environmental Equity: Reducing Risk for All Communities.* Volume II. Office of Policy, Planning, and Evaluation. Washington, D.C.: U.S. EPA.

———. 1994. *Environmental Justice 1994 Annual Report: Focusing on Environmental Protection for All People.* Office of Environmental Justice. Washington, D.C.: U.S. EPA.

———. 1995a. *Environmental Justice Strategy: Executive Order 12898.* EPA/200-R-95-002. Office of Environmental Justice. Washington, D.C.: U.S. EPA.

———. 1995b. *Permits Improvement Team National Stakeholder Report.* Office of Solid Waste and Emergency Response. Washington, D.C.: U.S. EPA.

———. 1996. *Concept Paper on Environmental Permitting and Task Force Recommendations.* Office of Solid Waste and Emergency Response. Washington, D.C.: U.S. EPA.

U.S. GAO (General Accounting Office). 1983. *Siting of Hazardous Waste Landfills and Their Correlation with Racial and Economic Status of Surrounding Communities.* Washington, D.C.: U.S. GAO.

U.S. GSA (General Services Administration). 1994–97. *Unified Agenda.* Washington, D.C.: Regulatory Information Service Center.

U.S. OMB (Office of Management and Budget). 1996. Fiscal Year Information Collection Budget (ICB). *Information Resources Management Plan of the Federal Government.* Washington, D.C.: Office of Information and Regulatory Affairs.

Vallely, Richard M. 1993. Public Policy for Reconnected Citizenship. In Helen Ingram and Steven Rathgeb Smith (eds.), *Public Policy for Democracy.* Washington, D.C.: Brookings Institution.

Vig, Norman J., and Michael E. Kraft, eds. 1994. *Environmental Policy in the 1990s.* Second edition. Washington D.C.: Congressional Quarterly Press.

Webster's 1988. *Webster's New World Dictionary of American English.* Third College Edition. New York: Webster's New World.

9

Comparison with Other Countries

How does the environmental regulatory system in the United States perform when compared with those in other advanced, capitalist countries? It is not possible to measure accurately and comprehensively the effectiveness of the U.S. system against those of other countries. Measurement is complicated by the fact that countries define, assess, and control pollution differently. It is, furthermore, challenging to gauge the degree to which laws actually affect pollution levels. Still, we can learn much about the U.S. system by examining comparable systems elsewhere.

Several factors make the comparison possible. To some degree, the respective control systems developed around the same time. Also, architects of environmental policy in each country have tended to borrow heavily from the U.S. model. (Future comparisons may become increasingly challenging as more countries move away from the media-specific U.S. model towards more integrated approaches.)

As a first rough gauge of performance, we examine pollution levels in thirteen advanced capitalist countries whose living standards, as measured by gross domestic product (GDP), are roughly similar to those of the United States (see Figure 1). We assume the countries have both the need and the resources to develop laws, institutions, and control techniques to tackle pollution problems associated with modern consumption patterns and production methods.

We then highlight key features of control systems in each country as a way to estimate whether laws and programs to administer them help to account for different pollution levels. We then provide an overview of efforts to integrate systems that have traditionally addressed pollution according to individual media. Finally, we briefly examine whether stringent pollution control policies in the United States contribute to trade imbalances.

POLLUTION LEVELS

Despite having one of the strictest pollution control regulatory systems in the world, the United States is still the largest polluter among our sample. In 1994, the last year for which the Organisation for Economic Co-operation and Development (OECD) environmental data are available, U.S. gross domestic product (GDP) was more than $6 trillion and the country's population was more than 260 million. At 9.8 million square kilometers, the United States ranks behind only Canada in terms of geographic size (OECD 1996, 256). The sheer economic, demographic, and geographic magnitude of the United States helps to explain why it has among the highest pollution emissions levels of countries in our sample. Even when we control for such factors, however, pollution levels in the United States remain high.

Air Emissions and Ambient Quality

To help compare levels of pollutants released to air among our sample countries, we examine several factors, including energy intensity, eco-

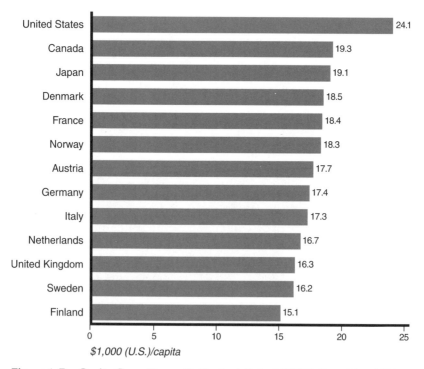

Figure 1. Per Capita Gross Domestic Product, Select OECD Countries, 1994.
Source: OECD 1996, 256–57.

nomic output, population, and the number of miles traveled each year by the average motorist. Data indicate that the United States has among the world's highest emissions of several pollutants. In terms of emissions per capita and GDP, the United States consistently ranks among the highest in emissions of sulfur oxides (SO_x) and nitrogen oxides (NO_x). Both substances stem primarily from industry and, to a lesser extent, the transportation sector. (OECD 1996, 77) Furthermore, the United States holds the dubious distinction of being the world's leading per capita contributor of carbon dioxide (CO_2) emissions. In 1993, the United States accounted for 24 percent of total world CO_2 emissions from energy consumption (OECD 1996, 77). Figure 2 shows CO_2 emissions per unit of GDP and per capita for select OECD countries.

CO_2 is among a group of "greenhouse gases" (including methane, nitrous oxide, chlorofluorocarbons and their substitutes, and other halo-

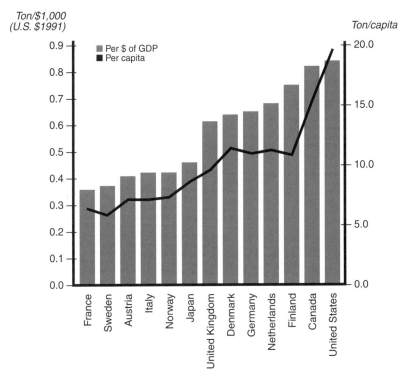

Figure 2. Carbon Dioxide Emissions of OECD Countries per Unit of GDP and per Capita, 1993.

Notes: Carbon dioxide from energy use only; international marine bunkers are excluded. GDP at 1991 prices and purchasing power parities.

Source: OECD 1996, 254–25.

genated compounds) which are thought to contribute to a gradual warming of the earth's temperature. Chlorofluorocarbons also are linked to the destruction of stratospheric ozone, which filters harmful light rays. The United States is the world's leading emitter of greenhouse gases, accounting for 17.8 percent of all emissions (OECD 1996, 154). CO_2 emissions from fossil fuel consumption account for roughly 80 percent of greenhouse gases emitted by the U.S (U.S. CEQ 1996, 208). In contrast, Japan, with the second-largest GDP, ranks fourth and is only responsible for 5 percent of global emissions.

Figure 3 shows that among the thirteen countries, the United States has the third-highest level of SO_x emissions per unit of GDP, behind the United Kingdom and Canada, and is second only to Canada on a per capita basis. High levels of smelting and refining of minerals and fossil fuel power generation contribute to Canada's high levels of SO_x and

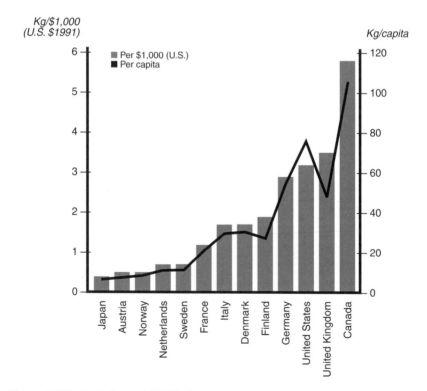

Figure 3. SO_x Emissions of OECD Countries per Unit of GDP and per Capita, 1993.

Note: GDP at 1991 prices and purchasing power parities.

Source: OECD 1996, 254–55.

other air pollutants (OECD 1995a, 96). In addition, cross-border air pollution from the United States contributes significantly to air pollution in southern Canada (OECD 1995a, 97). Japan, with the world's third-largest GDP per capita, has the lowest emissions of SO_x per capita and per unit of GDP of the thirteen countries analyzed. Japan has achieved a dramatic reduction in SO_x emissions since 1970 through several measures, including cleaner fuel, improved sulfur removal technology, and changes in the composition of the economy (OECD 1994d, 35).

Canada and Finland both have higher nitrogen oxides (NO_x) per unit of GDP than the United States (Figure 4). The high levels in Canada may be mostly due to the large number of Canadian industries involved in smelting and refining of minerals and chemical- and solvent-based manufacturing. In addition, Canada relies heavily on road transport (OECD 1995a, 96–97). On a per capita basis, however, the United States ranks

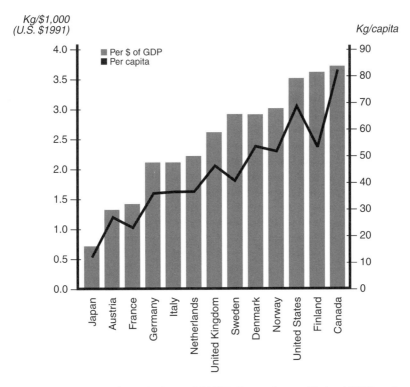

Figure 4. Nitrogen Oxide Emissions of OECD Countries per Unit of GDP and per Capita, 1993.

Note: GDP at 1991 prices and purchasing power parities.

Source: OECD 1996, 254–55.

Table 1. Status of Air Pollutants in the Top Ten Megacities, 1992.

City	SO_2	TSP	Pb	CO	NO_2	O_3
Tokyo/Yokohama	L	L	ND	L	L	S
Mexico City	S	S	M	S	M	S
São Paulo	L	M	L	M	M	S
New York	L	L	L	M	L	M
Shanghai	M	S	ND	ND	ND	ND
Calcutta	L	S	L	ND	L	ND
Buenos Aires	ND	M	L	ND	ND	ND
Seoul	S	S	L	L	L	L
Greater Bombay	L	S	L	L	L	ND
Rio de Janeiro	M	M	L	L	ND	ND
Los Angeles	L	M	L	M	M	S

Notes: TSP: total suspended particulates
S: Serious pollution, WHO guidelines exceeded by more than a factor of two.
M: Moderate to heavy pollution, WHO guidelines exceeded by up to a factor of two (short-term guidelines exceeded on a regular basis at certain times).
L: Low pollution, WHO guidelines normally met (short-term guidelines are exceeded occasionally).
ND: No data available or insufficient data for assessment.
Source: WHO/UNEP 1992, 8; 39.

above Finland. As with SO_x emissions, Japan is among the lowest emitters of NO_x of the OECD countries. Reductions in NO_x emissions in Japan between 1970 and 1989 are attributed to significant improvements in combustion technology, installation of denitrification units for exhaust gases, and use of three-way catalytic converters on passenger cars and light-duty trucks (OECD 1994d, 35).

Data with which to compare ambient air quality levels among the thirteen countries have not been developed. However, the World Health Organization (WHO) assessed the air quality conditions in twenty megacities (urban areas that have more than ten million people as of 1990 or are expected to have more than ten million people by 2000) (WRI 1994, 198). Table 1 shows the status of air pollution in Los Angeles and the top ten megacities, as ranked by 1990 metropolitan population figures. Although this table includes cities in less-developed countries, it is interesting to note that New York is among the leaders in the top ten megacities for meeting WHO standards, while Los Angeles, even though it had one of the smallest populations among the megacities in 1990, still fares poorly in comparison with several other megacities.

Extremely high numbers of vehicle miles traveled and number and size of vehicles on U.S. roads may be among the leading causes of air pollution. In the United States, transportation alone is responsible for 77 percent of total CO emissions, 45 percent of total NO_x emissions, and 31 per-

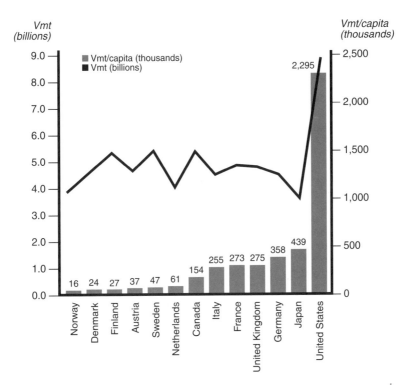

Figure 5. Vehicle Miles Traveled and per Capita, Select Countries, 1993.

Note: "Vehicle" refers to motor vehicles with four or more wheels, except for Japan and Italy, which include three-wheeled goods vehicles.

Source: OECD 1996, 256–57.

cent of total CO_2 emissions (OECD 1996, 77, 79). Figure 5 shows that while miles driven per person in 1993 range roughly between 1,000 and 1,500 in the other countries, the level in the United States is nearly 2,300.

The United States has the highest level of vehicle ownership in the world. At least 75 percent of the American population owns cars. Only a few other countries, including Canada and Italy, have ownership rates above 50 percent. In 1994, Americans owned approximately 192 million vehicles. Japan has the second-highest total number of vehicles, with approximately 60 million; however, both Japan and Italy count three-wheeled goods vehicles as motor vehicles (OECD 1995d, 218–19).

Measured in dollars, the United States has one of the most energy-intensive economies in the OECD. (See Figure 6.) The energy-intensity of a nation's economy also may be represented by its total energy use measured as tons of oil equivalent per unit of that nation's GDP. From this perspective, Canada and Finland both use more energy per unit of GDP

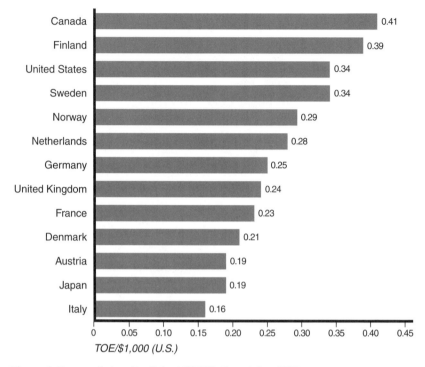

Figure 6. Energy Intensity, Select OECD Countries, 1993.

Note: TOE = tons of oil equivalent

Source: OECD 1996, 256–57.

than the United States. Canada's higher level of energy intensity is due primarily to its geographic size, cold climate, and energy consumption patterns (OECD 1995a, 137). The same factors probably account for low energy use levels in Italy and Japan.

Water Treatment Levels and Standards

Different collection, measurement, and reporting techniques make it impossible to compare water quality trends among countries. We present a few alternate, albeit imperfect, indicators of emissions from stationary and diffuse sources. Measures include wastewater treatment levels, drinking water compliance rates, and fertilizer and pesticide use patterns. Based on these indicators, the U.S. record on water pollution is slightly better than in the case of air pollution.

Among the thirteen countries, the United States provides wastewater treatment to more of its population than France, Canada, Italy, Norway, and Japan. About 72 percent of the U.S. population is served by treatment plants. The figure is low relative to levels in Denmark, the Netherlands and Sweden, whose rates are 98 percent, 97 percent, and 95 percent, respectively. Japan serves roughly 50 percent of population with waste-water treatment plants. While the level of U.S. population served is low relative to levels in most of the Scandinavian countries, it may represent close to the maximum percentage that makes economic sense, given dispersed U.S. population (OECD 1995a, 62; 1995b, 51; 1993c, 51).

Drinking Water

By 1990, most industrialized OECD countries provided drinking water that meets pollution standards to all of their urban and rural residents (WRI 1994, 275). However, a few countries serve less than 90 percent of the population with drinking water that meets their respective national standards. For example, while 100 percent of Japan's urban population has drinking water meeting national standards, only 85 percent of the rural Japanese population is served with drinking water that meets the same standards (WRI 1994, 274). In 1993, more than 90 percent of community water systems in the United States were in compliance with maximum contaminant limits (OECD 1996, 58).

Nonpoint Sources

The United States appears to be in a better position with respect to pollution from nonpoint sources than its counterparts, especially in the area of agriculture. Farmers in the United States use lower quantities of nitrogenous fertilizers and pesticides than all other countries examined except Canada. While farmers in the United States apply an average of 5.3 tons of fertilizer per square kilometer of arable, permanent cropland, farmers in the Netherlands use about 41.5 tons (OECD 1995b, 214). Dutch farmers apply roughly 1.8 tons of pesticides per square kilometer, while farmers in the United States use only about 0.2 tons (OECD 1995b, 156). The variation may be due to a number of factors, including the relatively smaller size of the other countries, high intensity crop mix, and intensive farming practices. For example, the Dutch flower bulb industry depends on intensive pesticide use.

Solid Waste

Different definitions of what constitutes municipal waste, as well as varying measurement techniques, make it difficult to compare precisely waste generation trends among nations. However, if nonhazardous solid waste

is understood as household and commercial refuse collected by municipalities, then the United States is one of the largest generators of municipal waste per capita in the world, as shown in Figure 7. A person living in
the United States generates about 730 kilograms of solid waste, while
someone in the United Kingdom generates less than half that amount
(OECD 1995b, 78).

One would expect U.S. numbers to be somewhat lower given that
many states in recent years have enacted their own recycling laws. By
1990 nearly forty states had enacted more than 140 recycling laws. These
state recycling laws have achieved approximately a 30 percent national
recycling rate for paper and a 20 percent recycling rate for glass (OECD
1996, 68). Nevertheless, the United States has among the lowest national
recycling rates for paper and glass (Figures 8 and 9). For example, while
the United States recycles about 20 percent of its glass, the Netherlands
recycles about 75 percent.

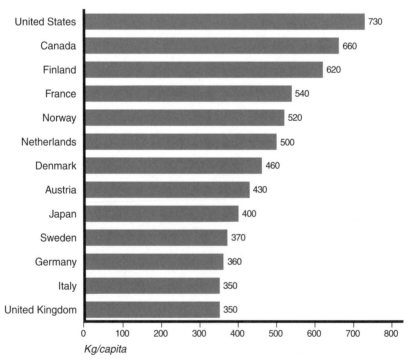

Figure 7. Per Capita Municipal Waste Generation, Select OECD Countries, 1993.

Note: France includes municipal sewage sludge (wet weight).

Source: OECD 1996, 254–55.

The inconsistency in the United States between waste reduction goals and actual per capita municipal waste generation may suggest that U.S. households are not persuaded to consume less and reduce waste. Possibly U.S. solid waste goals to reduce, recycle, and reuse waste are not ambitious enough, or perhaps states, which are largely charged with solid waste reduction, are simply failing to meet reduction targets. Finally, our nation's consumer-oriented culture may run contrary to an ethic of reducing waste.

Hazardous Waste Generation

Due to definitional ambiguities, it is difficult to compare hazardous waste generation trends. Because countries include industrial waste in their definitions of hazardous waste, it is difficult to tell whether all the wastes a

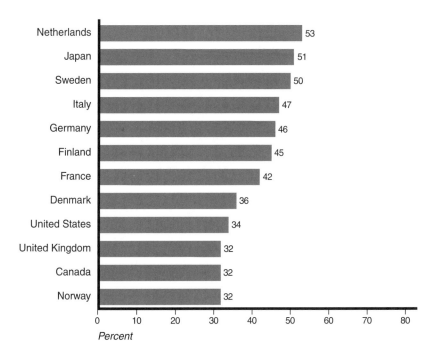

Figure 8. Waste Recycling Rates, Select OECD Countries, Paper and Cardboard, 1993.
Note: Data from Canada, Denmark, and Finland are for 1992; data for the Netherlands are for 1991. Recycling is defined as any reuse of material that diverts it from the waste stream, except for recycling within industrial plants and the reuse of material as fuel. The recycling rate is the ratio of the quantity recycled to the apparent consumption (domestic production + imports – exports).
Source: OECD 1995d, 171.

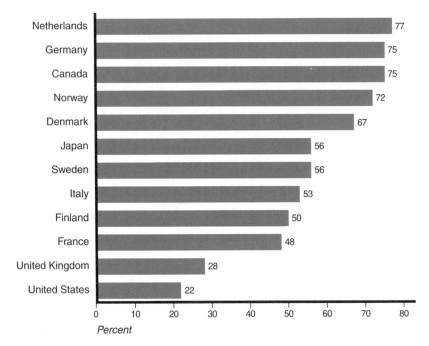

Figure 9. Waste Recycling Rates, Select OECD Countries, Glass, 1994.

Note: Data are for the most recent year available. For Italy, figure is percent of national production of glass containers for liquids. Tons collected as a percentage of national consumption, reusable glass bottled are excluded.

Source: OECD 1995d, 171.

particular country considers hazardous or industrial are included in available data. Figure 10 therefore defines industrial waste as waste from manufacturing industries (International Standard Industrial Code 3). This definition of hazardous waste includes waste from other industrial activities such as petroleum refining that are not manufacturing activities per se. Based on this measure, the United States is among the highest generators of industrial waste per unit of GDP. Finland, with the lowest per capita GDP of countries in our sample, has the highest hazardous waste level, at 206 kilograms per $1,000 U.S. dollars.

Based on the hazardous waste generation levels, one might expect to find large numbers of contaminated waste sites in the United States. Table 2 illustrates the number of sites estimated in several countries. The U.S. estimate is low compared with the Netherlands and Germany, but it is not possible to conclude that the United States has fewer contaminated sites than other countries. It may be simply that other countries define contamination more conservatively.

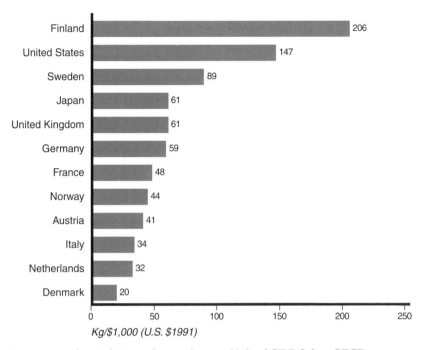

Kg/$1,000 (U.S. $1991)

Figure 10. Industrial Waste Generation per Unit of GDP, Select OECD Countries, 1993.

Note: Data are for the most recent year available; recent data for Canada are not available. GDP is at 1991 prices and purchasing power parities. The waste generated is from manufacturing industries (ISIC 3).

Source: OECD 1996, 254–55.

Toxic Emissions

Comparing toxics releases is difficult because countries employ different reporting methods. According to one report, the United States has among the highest levels of total toxic emissions (WRI 1994, 217). However, U.S. levels per unit of GDP are lower than a number of other OECD countries (as calculated by the authors from OECD 1995c, 124; WRI 1994, 217). Emissions per GDP in Germany top the list at 6,397 kilograms. Denmark had the lowest emissions per GDP, at 2,343 kilograms per $1,000. The United States is somewhere in between, with about 3,654 kilograms of toxic substances per $1,000 of GDP.

POLLUTION CONTROL PROVISIONS AND IMPLEMENTATION

Even when we adjust for factors such as living standards, population, and transportation distance, pollution emissions in the United States—

Table 2. Number of Sites and Funding Mechanisms for Select Hazardous Waste Cleanup Programs, Early 1990s.

Country	Estimated number of potentially hazardous waste sites	Sites slated for cleanup	Primary funding
United States[a]	38,000	1,320 (2,000–3,000 eventual size of NPL)	Business taxes and broad-based liability scheme for responsible parties
Canada[b]	4,800–10,000	40 orphan sites	Half from federal government and half from provinces/territories
Austria[c]	7,000; 300 need further assessment	40	Responsible parties voluntary cleanup; waste disposal and export taxes; federal matching funds
Norway[d]	2,696	450	Federal funds
Sweden[e]	8,000	250	Responsible parties; federal general revenues; municipal matching funds
Netherlands[f]	110,000	Not available	General revenues or responsible parties
Germany[g]	97,000–200,000	Not available	Responsible parties; waste taxes or fees; voluntary contributions; general revenues[h]
United Kingdom[i]	50,000–100,000	Not available	Federal grants/responsible parties

Notes: [a]Probst and others 1995, 17–18; 20. (In 1995, approximately 25,000 sites were removed from CERCLIS as they required no further action.)
[b]OECD 1995a, 86–87. (Responsible parties investigate and clean up their privately owned sites on their own initiatives.)
[c]Lohof 1991, 9.
[d]OECD 1993c, 57–58; 61. (In 1992, 747 sites were deemed to require no further action.)
[e]Lohof 1991, 31–32.
[f]OECD 1995b, 88–89. (Estimated number of sites includes industrial sites only, but other contaminated sites include landfills, car dumps and gasworks; landfills are cleaned up by the provinces.)
[g]Unless otherwise noted, OECD 1993a, 62.
[h]Lohof 1991, 13–14.
[i]Business Roundtable 1993, 19–20.

particularly for air, solid and hazardous waste—are persistently high. Below, we examine national standards and programs to control pollution in order to assess whether such controls or factors beyond the immediate scope of the regulatory system explain why U.S. emissions are higher than those in similar countries.

We first examine standards that specify how much air and water pollution is permissible in the environment, as well as standards that specify how pollution should be controlled. We then examine what types of administrative measures countries employ to meet these goals. Following air and water, we discuss what types of limits countries place on solid and hazardous waste, as well as what methods they use to control toxic substances. Finally, we compare different methods (such as recycling and pollution prevention) employed in each country to reduce waste and toxics.

CONTROL MEASURES: AIR AND WATER

Air Standards

Air pollution standards among developed countries are relatively similar. Of the thirteen countries examined, Germany's two-tiered standards most closely resemble those in the United States. The first tier seeks to prevent serious health hazards; the second is designed to prevent "considerable disadvantages and substantial impairments" (OECD 1993a, 35). Japan also has strict air quality standards.

In contrast, Canada does not have a law that sets national air quality standards. Instead Canada has three levels of national air quality objectives: maximum desirable (long-term goal); maximum acceptable (acceptable level of air quality, usually close to World Health Organization guidelines); and maximum tolerable (undesirable level) (OECD 1995a, 93). In Canada, provinces set air quality standards through their own environmental protection acts or specific air management legislation (OECD 1995a, 99).

Beyond such generalizations, comparisons are difficult because countries tend to employ different measurement techniques and monitoring technology. For example, Japan uses shorter averaging times than the United States and several other countries. Table 3 presents standards for the United States, Germany, Canada, and Japan. In most cases, allowable concentration levels appear relatively similar. Nitrogen dioxide and sulfur dioxide show the greatest variation.

Like the United States, several countries also require health- or technology-based standards to control air pollution (and treatment standards to control water pollution). For example, Canadian provinces issue site-specific permits that are based on ambient air quality objectives; in many cases, provincial regulations refer to the use of best available economic

Table 3. Ambient Air Quality Standards and Guidelines[a], Select Countries, Early 1990s.

Pollutant	Type of average	U.S. primary standards	Germany first tier standards	Japan	Canada national objectives (maximum acceptable)
Carbon monoxide	Annual mean				
	8-hour	9 ppm[b] (10 mg/m³)[b]	10 mg/m³	10 ppm (daily average of hourly values)	13 ppm
	1-hour	35 ppm[b] (40 mg/m³)[b]			31 ppm
Lead	Short-term (half-hour mean)		30 mg/m³		
	Annual mean		2 µg/m³		
	Maximum quarterly average	1.5 µg/m³			
	30-day			1.5 µg/m³ (30-day average)	
	24-hour				
Nitrogen dioxide	Annual mean	100 µg/m³	80 µg/m³	470 µg/m³ (daily average of hourly values)	53 ppb
	24-hour				106 ppb
	1-hour				213 ppb
	Short-term (half-hour mean)		20 µg/m³		
Ozone	Annual mean				15 ppb
	8-hour				
	1-hour			.06 ppm (average of hourly values)	82 ppb
	Maximum daily 1 hour average	235 µg/m³ [c]			
PM-10	Annual mean	50 µg/m³ [d]		100 µg/m³ (daily average of hourly values)	
	24-hour	150 µg/m³ [d]		200 µg/m³ (average of hourly values)	
Total suspended particulates	Annual mean		150 µg/m³		70 µg/m³
	Median of annual daily mean values				
	24-hour				120 µg/m³
	Short-term (half-hour mean)		300 µg/m³		

Sulfur dioxide				
Annual mean	80 μg/m³	50 μg/m³; 140 μg/m³ in winter	0.04 ppm (daily average of hourly values)	23 ppb
Median of annual daily mean values				
24-hour	365 μg/m³ [b]	40 μg/m³		
1-hour			0.1 (average of hourly values)	115 ppb
Short-term (half hour mean)				344 ppb
10-minute				

Notes: Particulates before 1987 were measured as TSP (total suspended particulates).

[a]Some standards or guidelines were converted by RFF for consistency in measurement units.

[b]Not to be exceeded more than once a year.

[c]The standard is attained when the expected number of days per calendar year with maximum hourly average concentrations above 0.12 ppm is equal to or less than 1, as determined according to Appendix H of the ozone NAAQS.

[d]Particulate standards use PM-10 (particles less than 10μ in diameter) as the indicator pollutant. The federal annual standard is attained when the expected annual arithmetic mean concentration is less than or equal to 50 μg/m³; the federal 24-hour standard is attained when the expected number of days per calendar year above 150 μg/m³ is equal to or less than 1, as determined according to Appendix K of the particulate matter NAAQS. The California standard is attained when the annual geometric mean concentration is less than or equal to 30 μg/m³.

[e]With ozone >0.10 ppm, one-hour mean or TSP >100 μg/m³, 24-hour mean.

[f]Ninety-eighth percentile of all daily mean values taken throughout the year.

[g]Guideline values for combined exposure to sulfur dioxide and suspended particulate matter (they may not apply to situations where only one of the components is present).

Sources: For United States: U.S. EPA 1994, 19; for California: WHO/UNEP 1992, Appendix 1, 222–28) and the Air Quality Management Plan, South Coast Air Quality Management District, Los Angeles; for Japan: OECD 1994d, 40; for Germany: OECD 1993a, 36; for Ontario: Standards Development Branch Ontario Ministry of the Environment, "Summary of Point of Impingement Standards, Ambient Air Quality Criteria (AAQC), and Approvals Screening Levels (ASL)," June 1994, Table 1 and fax from Shalini K. Venkataramaiah (Ontario Ministry of Environment and Energy) to Elise Annunziata, "Recommended Multimedia Standards and Guidelines for Lead," 12/8/95; for Canada: OECD 1995a, 94; for European Community: EEC Directives: 80/779/EEC Sulphur Dioxide and Particulates, 85/203/EEC Nitrogen Dioxide, 92/72/EEC Ozone Concentrations in the Air, 82/884/EEC Lead; for WHO: WHO/UNEP 1992, 12.

technology (OECD 1995a, 100). Under its Technical Instruction on Air Quality Control, Germany has adopted requirements that are similar to those in the United States. Germans set specific technology-based emission limits for numerous pollutants, require strict control technologies for new and existing industrial facilities, and set ambient air quality standards (OECD 1993a, 34). To promote control measures that employ continuously evolving technology, they tailor best available control technology requirements and emissions limits to individual industries. Unlike facilities in the United States, new and existing German plants must meet the same standards, but on different compliance schedules (OECD 1993a, 35).

Legislation in many of the thirteen countries does not spell out what kinds of technology firms must use; however, Japan, the Netherlands, and the United Kingdom increasingly promote the use of advanced or clean technology. For example, the Japanese have aggressively developed technologies to remove nitrogen from gas exhaust, and they also employ extensive energy efficiency measures in the industrial sector (OECD 1994d, 41). Japanese passenger car emission standards are among the most stringent in the world (OECD 1994d, 40). In comparison with California, Japanese standards by 1992 were twice as stringent for CO, slightly more stringent for hydrocarbons, and equivalent for NO_x (OECD 1994d, 43).

The United States was among the first countries to control emissions from mobile sources such as cars and trucks. EPA began to phase out leaded gasoline in the early 1970s and required three-way catalytic converters in all new cars beginning in 1981 (OECD 1996, 84). Gasoline in Japan was lead-free by 1987 (OECD 1994d, 37). The most significant reductions in the United States occurred in 1985 and 1986. The Clean Air Act, as amended, requires all automobile fuel to be completely lead-free by January 1, 1996 (Clean Air Act Amendments of 1990, Sec. 211(n)).

Italy, Norway, and most European nations only recently began to require the use of unleaded gasoline and catalytic converters. In 1989, Italian regulators took measures to reduce lead in gasoline. A 1991 EC Directive required that all new cars in the EC be equipped with catalytic converters; prior to the directive, only 0.1 percent of the motor vehicles in Italy had such devices (OECD 1994c, 129, 131). Norway adopted a standard equivalent to the 1983 U.S. standard requiring catalytic converters on all new cars in 1989 (OECD 1993c, 69).

While many countries employ economic instruments such as taxes to reduce pollution (OECD 1997), the U.S. emissions trading system is relatively unparallelled (OECD 1996, 85). One notable exception is Germany's bubble concept, which allows existing industrial facilities to meet one overall limit on each type of air emission rather than treat each stack or vent separately; that is, the facility is treated as if it had a bubble over it. Germany uses the bubble rule selectively, in part because the instrument was primarily created as a way to speed up retrofits of existing plants (OECD 1993a, 111).

One economic instrument commonly used in other countries but not in the United States is a pollution tax on automotive fuel. As a result, the price of automobile gasoline in other countries is generally much higher. Figure 11 compares gasoline prices and taxes among OECD countries. Both fuel price and fuel taxes are lowest in the United States. For example, in 1994 the U.S. tax on gasoline was about a third of its total price. In contrast, the Portuguese tax comprises around two-thirds of the price of gasoline. Most other countries have taxes that are at least 45 percent—and often up to 60 to 75 percent—of total gasoline prices (U.S. DOE 1994, 17).

Many countries also place an environmental tax on fossil fuels—an effort that the United States has debated, but failed to adopt. Germany was one of the first countries in Europe to use a significant tax differential for unleaded fuels (OECD 1993a, 142). Norway uses taxes as both a disincentive and an incentive mechanism. The government taxes the sulfur content of oil, but has also lowered taxes on unleaded gasoline to promote its use (OECD 1993c, 69).

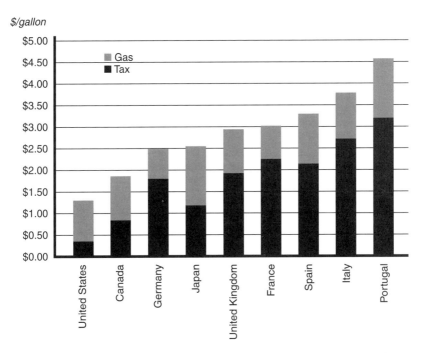

Figure 11. Unleaded Gasoline Prices and Tax Component, Selected Countries, 1994 (dollars per gallon).

Source: RFF calculations based on OECD 1995d, 230.

Several countries use revenues from fuel taxes to support environmental or conservation programs. The Netherlands collects an environmental tax on energy sources that are used as fuel, but not on those used as raw materials or feedstock. Between 1988 and 1992 the Dutch allocated revenue generated from this tax to finance government expenditures on environmental programs (OECD 1995c, 65).

Norway's CO_2 tax, introduced in 1991 and upgraded in 1992 and 1993, applies to a range of fuel combustion uses to encourage energy conservation and substitution (OECD 1993c, 69–70). Some Canadian provincial taxes are related to alternative fuels or vehicular efficiency (that is, taxes imposing a surcharge on less fuel-efficient vehicles or lower taxes on fuel-efficient vehicles). For example, Ontario places a graduated tax on new vehicles according to their fuel consumption. British Columbia, Manitoba, and Ontario impose lower taxes on vehicles that use alternative fuels (OECD 1995a, 100).

Water Standards

Comparing water standards among countries is even more challenging than comparing air standards because treatment levels tend to vary. For example, secondary treatment in one country may be comparable to tertiary treatment in another. Countries also tend to define water "use" differently.

The thirteen countries studied employ both ambient and technology-based standards to control water pollution. For example, Norway has a national standard that requires primary treatment of sewage, while additional sewage treatment levels are based on ambient quality guidelines for the receiving water. Norway bases its treatment levels on the size of the discharging system. According to the North Sea Declaration, Norway must provide secondary treatment of discharges from "large" municipal treatment plants that serve at least 5,000 people (OECD 1993c, 49).

Dutch technology-based standards for effluent discharges depend on the type of substance being discharged. For "blacklisted" substances such as mercury and cadmium, standards are met through the use of best technical means. For hazardous substances that are not blacklisted, best practicable means (not as strict as best technical means) apply and do not depend directly on the receiving water quality standards. For other non-hazardous substances, the receiving water quality standards are the main guide (OECD 1995b, 46). In the Netherlands, specific functions (uses) are assigned to water bodies; for each function there is an extensive set of quality standards for water and bottom sediments (OECD 1995b, 45–46).

Some countries discharge much of their wastewater untreated into receiving waters. For example, about half of Italy's municipal sewage effluents are discharged untreated into receiving waters, while one-quarter undergoes primary treatment only and the remaining quarter receives secondary treatment or higher (OECD 1994c, 85). Japan sets environmental quality standards and effluent reduction targets, but there are no deadlines or penalties for noncompliance. Thus, a large part of municipal and industrial wastewater is discharged without treatment (OECD 1994d, 67, 69).

Countries using an integrated approach to pollution control also tend to rely on technology-based standards. Thus, for example, the British integrated standard is Best Available Techniques Not Entailing Excessive Cost (BATNEEC). BATNEEC may be expressed in technological terms (that is, a prescribed technology) or in terms of release standards or performance standards. However BATNEEC is expressed, the inspectorate normally uses terms that will not inordinately limit development of cleaner technology or operators' choices of means to achieve prescribed standards (U.K. DOE 1993, 13). Identifying the optimal technology when considering all media has been a challenge, but in the absence of systematic multimedia criteria the British have required BATNEEC for air, water, and solid waste discharges.

Several water program features, such as the large capital expenditures on sewage treatment systems, may help to explain why the United States performs somewhat better than in the case of air pollution. The United States is one of a few countries that constructed an extensive municipal wastewater infrastructure (through use of heavy federal subsidies). From 1970 to 1990, the federal government contributed almost $60 billion in assistance—about 75 percent of the investment cost—for the construction of municipal sewage treatment works. In 1990, the State Revolving Fund Program replaced that program and provided federally subsidized loans to states to be used for municipal wastewater treatment projects and also for nonpoint source pollution programs (OECD 1996, 52).

Several developed countries have recognized the problem of pollution from nonpoint agricultural wastes and taken measures to reduce the use of fertilizers and pesticides. Some have adopted measures similar to best management practices used in the United States, such as monitoring systems, integrated crop management, relocation of animal confinement areas, revegetation, improved grazing systems, and improved irrigation management.

The United Kingdom's 1991 Code of Good Agricultural Practice for the Protection of Water outlines regulations for the management of farm wastes and the application of fertilizers and pesticides. In nitrate-sensitive areas, use of fertilizers and organic manure is restricted and farmers who voluntarily adopt measures to reduce nitrate leaching receive payments (OECD 1994e, 58). Similarly, Germans are using a variety of measures to stem pollution from diffuse sources, including programs to take some farmland temporarily out of production, and to promote less intensive farming practices. Municipal treatment plant operators also are installing denitrification facilities (OECD 1993a, 76).

Other countries use economic mechanisms to control pollution from diffuse sources and more generally to encourage water conservation. Norway, for example, has a national tax on the use of fertilizer at a rate of 19 percent on nitrogen and 11 percent on phosphorus. Some of the resulting revenue applies to the development of ways to reduce agricultural pollution (OECD 1993c, 52).

Few countries utilize water pricing systems to stimulate conservation; however, Japan provides an example of a progressive pricing system that is used to reduce consumption of drinking water (OECD 1994d, 74). Most countries, including the United States, have industrial wastewater disposal and treatment charges that are based on volume.

CONTROL MEASURES: WASTE AND TOXICS

U.S. requirements for handling solid and hazardous waste, as well as cleanup requirements for sites contaminated with hazardous waste, are among the world's most stringent. Furthermore, U.S. methods of assessing the potential harm posed to humans and the environment have historically been the most comprehensive.

Hazardous Waste

Most developed countries, including the United States, manage hazardous waste primarily through national laws that contain specific targets, disposal requirements, and restrictions on movement of hazardous waste. All storage, treatment, and disposal facilities that accept hazardous waste in the United States must have permits (Andrews 1992, 101). The United States prescribes stringent standards for landfills and incinerators that accept hazardous waste (40 CFR 264.300-317). Hazardous waste incinerators require permits and must comply with strict emissions limits (such as destruction and removal efficiency rates of over 99 percent) for metals and nonmetals (40 CFR 264.340-351).

Germany also has some of the strictest national disposal standards in the world (OECD 1993a, 57). Germany's main approach to risk reduction consists of requiring pretreatment and/or treatment of potentially hazardous wastes. Like the United States, Germany imposes very stringent design and monitoring requirements (such as double liners and leachate control) for all new land disposal sites (OECD 1993a, 56). Germany's extremely stringent requirements have led to very high prices for disposal. For instance, in 1990, landfill prices in western Germany for disposal of metal, oily sludges and asbestos were roughly double the median price of disposal in the rest of Europe (OECD 1993a, 54–55).

In the United States, all domestic shipments of hazardous waste must be documented in a detailed manifest that accompanies the waste through all stages of its disposal, sometimes referred to as a "cradle-to-grave" paper trail (OECD 1996, 67). Beyond documenting that hazardous wastes have been sent to permitted facilities, there is no restriction on where those wastes are sent (Andrews 1992, 101). Canadian federal law requires special training for vehicle operators, safe packaging and handling of hazardous materials, and consignment notes to accompany the wastes (OECD 1995a, 84–85). In addition to these minimum federal requirements, most Canadian provinces have their own waste registration and manifest systems for the transport of hazardous waste (OECD 1995a, 85). In contrast to the United States, some countries do not permit interstate waste shipments. For example, most wastes in Germany must be dealt with in the Länder (equivalent to a U.S. state) in which they were generated and not transported to other Länder (OECD 1995a, 57).

The United States exports a small amount of hazardous waste to Canada and Mexico for disposal (OECD 1996, 72). Germany and Italy also export hazardous waste for disposal (OECD 1993a, 57; OECD 1994c, 71). In contrast, the United Kingdom imports hazardous waste for treatment and disposal. Between 1992 and 1993, the U.K. imported 46,000 tons of hazardous waste, mainly from twenty-three western European countries (OECD 1994e, 73). The United Kingdom does not export any hazardous waste, although its export of nonferrous metal for recycling has been criticized (OECD 1994e, 73).

Solid Waste

While Americans generate more solid waste than their economic counterparts, the United States has among the most stringent municipal solid waste disposal requirements. Municipal solid waste in the United States and other countries is either landfilled or incinerated. U.S. waste law prescribes minimum national requirements for municipal waste landfills and combustion. Municipal solid waste landfills do not need to be permitted, but they must meet location restrictions, operating and design criteria (such as groundwater and methane gas monitoring), closure requirements, and financial assurance criteria. In addition to these requirements, new landfills must have a liner and a leachate collection system (40 CFR 258). Municipal waste burners are not regulated under RCRA, but the U.S. Clean Air Act sets limits for combustion emissions, including lead, mercury, and dioxin (Clean Air Act Amendments Sec. 129). In addition, municipal solid waste combustion ash is regulated as hazardous waste, according to a recent Supreme Court determination (*City of Chicago, et al., v. Environmental Defense Fund, et al.* 114 S.Ct. 1588 [1994]).

A contrasting example is Canada, where few landfills meet guidelines for protective linings and leachate collection, and most sites are rudimentary (OECD 1995a, 83–84). In the United Kingdom, where land space is sufficient and landfill is the primary waste disposal option, containment by clay or artificial liner has only been introduced in the last decade and is not common practice (OECD 1994e, 70–71). In some countries where land space is scarce, incineration of solid waste may be preferred. For example, Japan incinerates 75 percent of its municipal solid waste, but, unlike the United States, Japan has no national regulations to control emissions of dioxin, furans, or heavy metals like mercury from incineration (OECD 1994d, 52–53).

Because U.S. law implicitly encourages states to develop and implement their own waste management programs, there is less federal government involvement in setting recycling goals and targets than occurs in most other countries. For example, in the Netherlands, the public sector is required to provide treatment and disposal facilities not only for municipal waste but also for waste from industry, except for hazardous waste. Construction and operation of these facilities is mostly carried out by public-private partnership (OECD 1995b, 81).

Differences among countries in generation rates may be due in part to incentives to reduce and recycle solid waste. While many states have passed recycling legislation, the United States does not have national recycling laws or programs. Most of the countries in our sample support national solid waste recycling programs. For example, Norway has a fee on all nonreturnable beverage containers, a deposit-refund fee on reusable glass and plastic beverage containers, and a deposit fee on all cars sold (OECD 1993c, 60). To support recycling of certain waste products, Italy has a charge on lead batteries and used oil, as well as a tax on plastic shopping bags (the latter tax was abolished in 1993 and replaced with a tax on virgin polyethylene film) (OECD 1994c, 70). Japan's recycling law promotes the use of "designated byproducts," such as slag from iron and steel operations; coal ash from electric utilities; and earth, sand, concrete blocks, asphalt and timber from the construction industry. Designated byproducts, which are eligible for recycling, account for over 40 percent of industrial waste. In Japan, the waste generators are responsible for preparing plans to increase the recovery and use of industrial byproducts (OECD 1994d, 60).

In 1988, the Dutch Memorandum on the Prevention and Recycling of Waste set quantitative targets for prevention, recycling, and disposal for twenty-nine (now thirty) priority waste streams by the year 2000. Additionally, the Dutch National Environmental Policy Plan (NEPP) and

NEPP Plus set goals for all types of waste. The targets for 2000 include a 10 percent prevention target (assuming a 2 percent annual increase in economic growth), increased recycling, and a sharp decrease of landfill disposal (OECD 1995b, 79).

The United Kingdom's Environmental Protection Act requires local collection authorities to develop recycling plans. The government has set a target of 50 percent by 2000 for recycling of household waste. In addition, the government is consulting with industries to help them set voluntary recycling goals (OECD 1994e, 74–75).

As in the case of recycling, there are no national packaging reduction laws in the United States. Some countries, including Netherlands, Germany, and Canada, have adopted national policies to reduce the amount of packaging and promote reusable packaging. The Dutch adopted a voluntary plan that by the year 2000 aims to reduce the total amount of packaging to at least 1986 levels by promoting the adoption of reusable packaging (OECD 1995b, 86). Germany's Packaging Ordinance of 1991 requires that waste generators take back wastes for recovery or proper disposal. The ordinance also reduces packaging by charging the generator fees. For example, generators are charged an extra DM 0.5 (approximately $0.36) for each single-use beverage container that is 0.2 liters (roughly half-pint) or more. Up to eight million tons of packaging are targeted for recycling and recovery operations annually (OECD 1993a, 59–60).

In Canada, the National Packaging Protocol sets targets for reducing packaging wastes and requires provincial and municipal governments to develop, in cooperation with industry, plans to reduce and recycle waste packaging (OECD 1995a, 80). Canada has two other national waste minimization programs; one encourages voluntary reduction of hazardous waste and the other provides data on the supply and demand of industrial waste for recycling (OECD 1995a, 81).

Contaminated Sites

Little information is available concerning other countries' remedial action processes and requirements, but some countries follow Dutch cleanup standards or negotiate cleanup requirements. The Dutch standards are numerical criteria modified by site conditions. The "Dutch list" establishes three levels of contamination and required activity: Level One is a reference for clean media, Level Two triggers further investigation, and Level Three triggers remediation. Countries such as the United Kingdom have a less formal process, negotiating standards on a case-by-case basis. Some developed nations, such as Japan, have not established any kind of remedial action process with cleanup standards or goals (Business Roundtable 1993, 19).

European methods of selecting sites for cleanup and determining how clean a site should be depart significantly from those in the United States. While some countries clean up the worst sites first, U.S. laws call for the cleanup of hazardous waste sites that pose immediate threats to human health and the environment. Apart from emergency situations, the only priority-setting for cleanup is that a hazardous waste site be listed on a National Priorities List. In contrast, all sites slated for investigation and cleanup in the Netherlands are classified in ten categories according to urgency, and a timetable is established for investigation and cleanup of each class (OECD 1995b, 89). Austria classifies sites that have highest priority as "Class I"; requests for funding from the national government must be submitted within one year of an official site classification decision (Lohof 1991, 9–10). Canada also has a prioritization system—high, medium, and low—for cleanup of its "orphan sites" (sites with no existing owner) (Business Roundtable 1993, 19). (See Table 2.)

In addition to having different prioritization strategies, the U.S. system tends to cast a much wider net of liability. While most countries impose a liability scheme for collecting cleanup costs from sites contaminated with hazardous waste, CERCLA stipulates a strict, retroactive, and joint and several liability scheme for the cleanup of U.S. hazardous waste sites (Probst and others 1995, 13). As a result, the process in the United States is typically characterized as more contentious and cumbersome. CERCLA holds multiple parties responsible for waste disposal acts that were legal at the time of contamination (Probst and others 1995, 14); almost anyone connected with the site—disposers, owners, generators, and transporters—can be held liable for the entire cost of cleanup (Lohof 1991, 4). In other countries, owners or operators of the contaminated sites are usually held responsible for contamination. Generators who legally ship to an approved disposal facility are seldom held accountable for cleanup costs (Business Roundtable 1993, 3); multiparty sites are rare outside the United States. In Austria, joint-and-several cleanup liability exists, but it has not played a significant role because of the absence of multiparty sites (Lohof 1991, 9).

Hazardous waste site cleanup efforts in other countries rarely involve litigation. The Dutch government in 1992 concluded a voluntary agreement that provided for the investigation and cleanup by industry of the 110,000 identified industrial sites. Companies that own or use the sites are responsible for these activities, and must pay for them. The Dutch also give a prior tax deduction to private companies for the cost of planned cleanup work. The Dutch Soil Protection Act enacted in 1994 authorized provinces to order polluters, owners, or users to clean up sites, and made voluntary cleanup work subject to prior notification to and approval by provinces (OECD 1995b, 89).

Austria's program for hazardous waste cleanup is another example of public sector involvement and voluntary cleanups. Responsible parties who offered to participate in cleanups were eligible to apply before December 1992 for matching funds from the Austrian federal government. The government's site assessment and cleanup fund was to be used to finance the cleanup of orphan sites as well as the government's share of matching-fund cleanup projects carried out with the cooperation of industry. Financial assistance was granted only for sites that were reported to state authorities by the cutoff date. The fund, financed by taxes on the disposal and export of all types of hazardous and nonhazardous wastes, paid, on average, 50 percent of the cleanup costs (Lohof 1991, 9–10).

Canada's cleanup strategy is unique. The National Contaminated Site Remediation Project (NCSRP), jointly funded by the federal government and the provinces or territories, was established to pay for the entire cost of forty orphan sites (OECD 1995a, 87). The project may recover remediation costs if a viable responsible party is identified later. As of 1993, the Canadian government had not initiated any cost recovery actions under the NCSRP (Business Roundtable 1993, 9).

Toxics

It is impossible to say why toxic emissions rates vary, but it is possible to describe differences in the way countries decide which chemicals to regulate. While some notable differences exist between risk assessment methods in the United States and its economic counterparts, evaluation methods are becoming more uniform. Regulators in the United States have attempted to standardize chemical assessment procedures through a formal, four-stage model that uses mathematical extrapolation methods to estimate the effects of cancer-causing chemicals. Methods abroad still tend to be more informal, with decisions made by panels of scientific experts. Trade and territorial alliances, such as European unification, are the primary factors behind increased harmonization. However, international organizations such as the World Health Organization have long sought to standardize chemical testing protocols (Mazurek 1996).

A notable difference between the United States and its European counterparts is the way laws determine which new chemicals to test. In 1993, an EC directive specified testing, notification, and control procedures for existing chemicals. Further, it required chemical manufacturers to report health and safety data to the European Union for chemicals produced in amounts greater than 1,000 tons per year. From these data, the European Union established a priority chemicals list for testing that is based on risk and volume (Mazurek 1996, 28). The EC system further imposes three tiers of test requirements on new chemicals: more stringent tests are applied to chemicals produced in higher volumes. In contrast, the U.S. system has no testing requirements and most new chemical notifications do not contain any toxicity test data.

Since 1986, the United States has employed information as a tool to help firms and interested citizens identify what types of toxics firms use and transport. EPA developed the Toxics Release Inventory in response to Section 313 of the Superfund Reauthorization Amendments. EPA collects data annually from firms and makes reports available to the public (Chapter 5); the European Community is now preparing to institute a similar system throughout Europe (EC 1996).

INTEGRATED POLLUTION CONTROL

So far, we have examined pollution levels and programs to control pollution according to the medium in which pollutants are released. However, countries in Europe and elsewhere are attempting to move beyond fragmented approaches to pollution control. (For an excellent overview, see Hersh 1996.)

In general, integrated pollution control aims to prevent pollution and increase protection of air, water, and land by making holistic decisions that address cross-media problems resulting from pollutant discharges to the environment. The United Kingdom, Sweden, Norway, Denmark, France, the Netherlands, and the European Union have all adopted some form of integrated pollution control. The U.K. and Swedish variants primarily use multimedia permits to reduce impacts to the environment and promote the adoption of cleaner processes. The Dutch system is more comprehensive; the National Environmental Policy Plan sets out an ambitious set of goals to reduce pollution across the economy, including households, industry, and agriculture.

The United Kingdom's Environmental Protection Act of 1990 established an integrated pollution control system and a single agency to administer the program. Her Majesty's Inspectorate of Pollution was created from extant agencies charged with controlling air pollution, hazardous waste, and other environmental problems. Integrated pollution control requirements target industries with the greatest potential for serious pollution (U.K. DOE 1993, 29). The regulated substances are subject to special requirements that specify use of best available techniques not entailing excessive cost to prevent release to air, water, or land of the substances (U.K. DOE 1993, 4). If prevention of such releases is not practicable, then targeted firms must use the techniques to minimize releases and render them harmless (U.K. DOE 1993, 4–5).

In most respects, the system appears to operate through the same mechanisms as traditional permitting in the United States: operators of regulated processes must apply for integrated pollution control authorization from the unified inspectorate before operating a new process or substantially modifying an existing one (U.K. DOE 1993, 5). The inspectorate must make the application available to the public for examination and comment (U.K. DOE 1993, 8). The distinction is that the specified techniques promote the use of the "best practicable environmental option," the combination of approaches that minimizes pollution in *all* media of the environment (U.K. DOE 1993, 10). It is still too early to tell how well the integrated British system will succeed in meeting its goals.

The Commission of the European Communities (CEC) borrowed from the U.K. system to develop a council directive on integrated pollution prevention and control. The directive would apply integrated pollution control to certain defined installations, or processes, that have a "high potential for pollution." It emphasizes that integrated pollution control is not just a way to balance the decision options regarding emissions to three media but also encourages pollution minimization (CEC 1994, 13–14).

The CEC directive requires the use of best available technology to set emissions limit values and to prevent emissions to the environment from certain installations, where "practicable." Where emissions prevention is not practicable, the goals of integrated pollution control are to minimize emissions and to promote energy and raw material efficiency. The directive is not intended to require specific technologies or techniques (CEC 1994, 13–14).

The U.K. and E.U. integration initiatives are focused primarily on industrial processes. Several OECD countries have adopted an integrated approach to pollution control as a way of achieving the broader goal of sustainable development. The Netherlands, for example, developed the National Environmental Policy Plan in 1989, with a goal of sustainable development by 2010 (Netherlands 1994, 5). The plan described environmental conditions in the Netherlands and grouped the main environmental problems and pollutants according to the economic activities responsible. Environmental problems were classified into one of eight environmental themes—climate change, acidification, eutrophication, dispersion, waste disposal, local nuisance, water depletion, and resource management (Netherlands 1994, 6). In order to reach the goal of sustainability by 2010, the government sets quality objectives to be achieved by 2010 for each theme and defines the percentage reductions in key pollutants required to meet the objectives. For example, a sustainable level of acidification would require a 70 to 80 percent reduction in acid emissions from 1985 levels by 2010 (Netherlands 1994, 8).

Target sectors, including agriculture, traffic and transport, industry and refineries, gas and electricity supplies, construction, and consumers and retail trade, are responsible for achieving emissions reduction targets for specified substances and waste streams (Netherlands 1994, 8). To meet emissions reduction targets, each sector must negotiate with the central government, local governments, and other sectors (Netherlands 1994, 17).

To meet such targets, the Dutch use covenants. Industry representatives or local authorities and the central government may enter into covenants as a means to speed up environmental improvements while legislation is still pending (OECD 1995b, 130). Covenants contain provisional requirements that are legally enforceable in civil courts. Approximately twenty-six environmental covenants have been agreed to by industry and the central government, dealing with products, packaging, waste, and emissions (Hersh 1996; OECD 1995b, 131).

Other Dutch legislation and policy have subsequently been developed to meet the objectives of sustainable development and integrated environmental policy planning. A second plan was developed in 1993 to formalize a long-term action strategy to meet the earlier objectives and targets (Netherlands 1994, 22). Another action, the Environmental Management Act of 1993, effectively simplified Dutch legislation by replacing numerous sectoral laws with a single act. The act also includes an integrated permitting system that will enable installations to apply for a single permit covering all operations (Netherlands 1994, 17).

Like the Netherlands, Norway has adopted quantitative emissions reduction targets to achieve sustainable development (OECD 1993c, 79). The Norwegian Pollution Control Act, as amended in 1993, covers all forms of pollution to air, water, and land. The act, based on integrated pollution control, focuses on, among other things, pollution prevention, cost-effective solutions, and phaseout of hazardous substances or use of best available technology. Norway also employs a system of cross-media emissions permits (OECD 1993c, 25–26).

Like Norway, Sweden has a single Environmental Protection Act that regulates the discharge of pollutants released to air, water, and land. The Swedish system of controlling pollution is based on "best practical means." Rather than issuing national pollution standards for air and water, as is done in the United States and many other countries, Sweden uses an integrated permitting and inspection system that regulates the total emissions from each plant or factory (Hinrichsen 1990, 147–48). By mandating operating permits and controlled inspections for polluting industries on a plant-by-plant basis, Swedish authorities are able to regulate emissions to air, water, and land as parts of a whole package (Hinrichsen 1990, 148).

INTERNATIONAL COMPETITIVENESS AND TRADE

Do stringent environmental regulations harm the competitiveness of U.S. firms? The question is a source of intense debate. The conventional view is that environmental regulations hinder U.S. competitiveness in international markets by imposing greater costs and restrictions on industries in the United States than are imposed on competitors in other countries. Alternatively, a newer group of "revisionists" holds that regulations drive firms and the economy as a whole to become more competitive in international markets (Jaffe and others 1995, 133). Jaffe's review of the economics literature regarding this debate concluded that while some of the foundations of the conventional view can be disputed, neither the conventional nor the revisionist view is supported by definitive evidence (Jaffe and others 1995, 159).

The conventional argument is that U.S. regulations put American firms at a competitive disadvantage. Conventionalists claim that regulations limit industry's choice of technologies, product design and mix, plant location, and other important production decisions. They argue that firms must allocate investment and operating budgets to reduce environmental impacts, even if these expenditures will not be recovered. In addition to direct compliance costs, industries face delays and uncertainties in dealing with regulatory requirements (Repetto 1995, 1). According to conventionalists, other industrialized countries have adopted tight requirements and stringent legislation to address pollution control that often appear as stringent as U.S. requirements; however, enforcement and application of such requirements in other countries are often not comparable to practices in this country (Quarles 1995, 18).

Aside from the stringency of environmental regulations, the form of these rules may affect business investment decisions. For example, some U.S. environmental regulations go beyond mandating numerical standards for discharges and require specific control technologies and processes (Jaffe and others 1995, 135). Conventionalists assert that the prescribed technological controls and standards that must be achieved by U.S. industrial operations are usually less flexible and more demanding than in other countries. Thus, the burdens of satisfying regulatory requirements are far more severe in the United States than in other industrialized countries (Quarles 1995, 19). Conventionalists believe that countries that avoid mandating specific technologies and allow new, innovative technologies to comply with discharge standards have a distinct competitive advantage (Jaffe and others 1995, 135).

Another burden on U.S. firms is the amount of paperwork and the nature of the relationship between the regulator and regulated. Quarles, in his analysis of environmental regulation for Pfizer Inc., says U.S. regulations are set within a framework of governmental supervision that is "more onerous" and accompanied by far more extensive monitoring, recordkeeping, and reporting requirements than in other competing countries (Quarles 1995, 19). Jaffe and others (1995, 135) agree that regulatory decision-making in the United States is often very time-consuming, involves extended public involvement, and is characterized by litigation and other legal challenges. Quarles (1995, 18) says that cooperation and compromise between government and industry is much more common in western Europe and Japan than in the United States.

Porter first challenged the conventional view in *The Competitive Advantage of Nations* (1990). A more recent article by Porter and van der Linde synthesizes these ideas. Porter disputes the notion that environmental regulation necessarily conflicts with competitiveness; instead, he claims that well-designed, strict environmental regulations can be a catalyst for innovation that will increase companies' profits by lowering product costs (Porter and van der Linde 1995, 120). In Porter's view, innovation will allow companies to use their production inputs (raw materials, energy, and labor) more efficiently and thus offset the costs of compliance with environmental regulations. Ultimately, he claims, this "resource productivity" will make companies more competitive, not less (Porter and van der Linde 1995, 120).

With respect to foreign competition, Porter believes that the United States should develop strict environmental standards at the same time as or slightly ahead of other countries. By doing so, the United States can minimize any possible competitive disadvantage associated with foreign companies that are not subject to the same standards. By developing regulations slightly ahead of foreign competitors, the United States may maximize export potential in the pollution control market by raising incentives for innovation. In other words, when the U.S. standards lead those of its competitors, domestic companies have "early-mover" advantages (Porter and van der Linde 1995, 124). As an example, Porter and van der Linde point out that because Germany adopted recycling standards earlier than most other countries, German companies had early advantages in developing less packaging-intensive products that cost less and are in demand in the marketplace (Porter and van der Linde 1995, 127).

Porter cautions that indiscriminate application of tougher environmental regulations will not increase competitiveness. Examples of regulations and policies that promote competitiveness include pollution prevention, market incentives, and performance-based regulation, rather than prescriptive calls for "best available technology" (Porter and van der Linde 1995, 124).

Repetto also maintains that the conventional view is incorrect, but he argues that it is so because it uses flawed models to evaluate the relationship between regulation and competitiveness. Current models employ estimates of productivity growth that measure only the effects of regulation on industry costs but fail to account for the reductions in pollution damage attributable to those regulations. Repetto argues that productivity measurements that include both the costs and benefits of environmental regulations may lead to the conclusion that environmental regulations raise the rate of productivity growth (Repetto 1995, 5). Repetto also suggests that while some regulations may affect one type of domestic industry more than another, the "effects of environmental regulation on trade shouldn't be judged at the level of the individual firm or even the individual industry" but rather at the level of the economy as a whole (Repetto 1995, 4–5). Finally, Repetto says that there is little evidence that environmental regulations have adverse effects on competitiveness, claiming: "Contrary to common perceptions, higher environmental standards in developed countries have not tended to lower their international competitiveness. There has been little systematic relationship between higher environmental standards and competitiveness in environmentally sensitive goods (those that incurred the highest pollution abatement and control costs...)" (Sorsa 1994, i).

Indeed, according to one World Bank study, industries heavily affected by environmental regulation did relatively well in international trade (Sorsa 1994, i). Between 1970 and 1990, the industrial countries' overall share in world exports declined from 74.3 percent to 72.7 percent, mainly because the rest of the world experienced faster economic growth and now contributes a larger share of world output than before. The industrialized countries' share of world exports of manufactured goods declined even more (from 91.3 percent to 81.3 percent), largely because expenditures and output in wealthy industrialized countries shifted away from manufacturing toward services, while in developing countries they shifted from agriculture toward manufacturing. However, within the category of manufactured exports, industrialized countries' share of exports in industries that experience the highest pollution control costs declined very little (from 81.3 percent to 81.1 percent). The sectors in which the industrialized countries lost their comparative advantage, such as textiles, apparel, footwear, and other light manufactures, were not those heavily affected by environmental regulations, but instead were those in which labor costs are a large fraction of total costs (Repetto 1995, 6–7). Repetto says that these trends do not necessarily suggest a causal relationship between trade and regulatory compliance costs. They simply show that other, larger forces are affecting the world economy and that there is no consistent link between environmental regulation and trade performance (Repetto 1995, 7).

Repetto (1995, 8) also shows that U.S. investment flows to developing countries do not support the contention that multinational companies are relocating environmentally sensitive industries to countries that are less regulated. Of the total U.S. direct investment abroad in pollution-intensive industries, 84 percent ($5.6 billion) went to other developed countries. To the extent that developed countries are exporting their polluting industries, they are sending them to each other rather than to the developing countries and are dealing with comparable environmental standards and regulations.

Jaffe and others (1995, 157–58) conclude that "overall, there is relatively little evidence to support the hypothesis that environmental regulations have had a large adverse affect on competitiveness." While the long-run social costs of environmental regulations may be significant (including adverse effects on productivity), "studies attempting to measure the effect of environmental regulation on net exports, overall trade flows, and plant-location decisions have produced estimates that are either small, statistically insignificant, or not robust to tests of model specification."

Jaffe and others (1995, 159) found little empirical evidence in support of the conventional hypothesis that environmental regulations hurt competitiveness, but they also found "little or no evidence supporting the revisionist hypothesis that environmental regulation stimulates innovation and improved international competitiveness." They suggest that the truth in the debate about the relationship between environmental regulations and competitiveness lies somewhere in between.

SUMMARY

No attempt to measure the overall effectiveness of the U.S. system vis-à-vis other countries has been made in this chapter. The United States is likely ahead in some areas and behind in others, while in many (perhaps most) areas, direct comparison is simply not possible.

There are a few instances where the United States has among the lowest levels of pollution. The United States uses fewer agricultural chemicals, which contribute significantly to nonpoint source pollution, than the OECD countries examined. For both nitrogenous fertilizer and pesticide use, the United States has the second-lowest level per square mile of arable and permanent cropland. World Bank data also suggest that the United States ranks comparatively low in annual toxic releases per unit of GDP.

Nevertheless, levels of air pollution and solid waste generated usually rank higher per capita and per dollar of GDP than levels in most other developed countries. For example, the United States is the leader in per capita emissions of NO_x and CO_2. U.S. emissions of these air pollutants per dollar of GDP are either tied for the highest or second-highest among the countries we examined. U.S. emissions of SO_x per capita and per unit of GDP are second only to Canada. The United States also has the highest per capita level of municipal waste generation, as well as the highest level of industrial waste generation, per unit of GDP.

Compared with the pollution control systems of several other developed OECD countries, U.S. air standards are among the most stringent, and U.S. enforcement is at least as effective. It is therefore likely that persistently high levels of air pollution in the United States are driven by factors outside the pollution control system. For example, the high volume of road traffic and the high number of vehicles per capita in the United States are most likely due to the high per capita income, low fuel prices, driving practices (such as high vehicle miles traveled for home-to-work commutes), and the enormous size of the country. Also, high U.S. energy use may be due to geographic size and long driving distances, as well as larger houses and other energy consumption patterns. Thus, instruments such as energy taxes may be needed to meet emissions reduction goals.

More so than the systems of many of the countries examined in this chapter, the U.S. pollution control regulatory system employs tools to control pollution after it has been created, rather than measures to decrease the amount of pollution produced. In contrast, other countries are moving away from individual approaches toward more integrated systems. As a result, the United States, once considered the leader in environmental institutions, threatens to become a laggard.

Finally, the United States also seems to rely more heavily on litigation as a means to enforce its regulatory system than the other nations examined. While economic, political, and cultural peculiarities influence the character of the U.S. system, it may be that more cooperative, less adversarial approaches can better promote pollution control.

REFERENCES

Andrews, Richard N. L. 1992. Environmental Policy-Making in the United States. In *Towards A Transatlantic Environmental Policy*. Washington, D.C.: The European Institute.

Business Roundtable. 1993. *Comparison of Superfund with Programs in Other Countries*. Washington, D.C.: Business Roundtable.

CEC (Commission of the European Communities). 1994. *Proposal for a Council Directive on Integrated Pollution Prevention and Control*. COM(93) 423 Final. Brussels: CEC.

55455455

EC (European Community). 1996. Council Directive 96/61/EC, 24 September, Articles 15 and 19.

Hersh, Robert. 1996. *A Review of Integrated Pollution Control Efforts in Selected Countries.* Discussion Paper 97-15. Washington, D.C.: Resources for the Future.

Hinrichsen, Don. 1990. Integrated Permitting and Inspection in Sweden. In Irwin and Haigh (eds.), *Integrated Pollution Control in Europe and North America.* Washington, D.C.: The Conservation Foundation..

Hunter, Roszell D. 1991. Proposed Waste Transfer System within the European Community. *BNA International Environment Reporter:* 14(25): 695–703.

Jaffe, Adam B., Steven R. Peterson, Paul R. Portney, and Robert N. Stavins. 1995. Environmental Regulation and the Competitiveness of U.S. Manufacturing: What Does the Evidence Tell Us? *Journal of Economic Literature* 33: 132–63.

Lohof, Andrew. 1991. *The Cleanup of Inactive Hazardous Waste Sites in Selected Industrialized Countries.* Discussion Paper 069. Washington, D.C.: American Petroleum Institute.

Mazurek, Janice. 1996. *The Role of Health Risk Assessment and Cost-Benefit Analysis in Environmental Decision Making in Selected Countries: An Initial Survey.* Discussion Paper 96–36. Washington, D.C.: Resources for the Future.

Netherlands. 1994. *Towards a Sustainable Netherlands: Environmental Policy Development and Implementation.* The Hague: Ministry of Housing, Spatial Planning and the Environment.

———. 1995. *Environmental Program of the Netherlands 1986–1990.* The Hague. Ministry of Housing, Physical Planning and Environment, Ministry of Agriculture and Fisheries, Ministry of Transport and Water Management.

Nishimura, H., ed. 1989. *How to Conquer Air Pollution: A Japanese Experience.* New York: Elsevier Science.

OECD (Organisation for Economic Co-Operation and Development). 1989. *OECD Environmental Data Compendium 1989.* Paris: OECD.

———. 1991. *OECD Environmental Data Compendium 1991.* Paris: OECD.

———. 1993a. *Environmental Performance Reviews: Germany.* Paris: OECD.

———. 1993b. *Environmental Performance Reviews: Iceland.* Paris: OECD.

———. 1993c. *Environmental Performance Reviews: Norway.* Paris: OECD.

———. 1993d. *Environmental Performance Reviews: Portugal.* Paris: OECD.

———. 1993e. *OECD Environmental Data 1993.* Paris: OECD.

———. 1994a. *Applying Economic Instruments to Environmental Policies in OECD and Dynamic Non-Member Economies.* Paris: OECD.

———. 1994b. *Environmental Indicators: OECD Core Set.* Paris: OECD.

———. 1994c. *Environmental Performance Reviews: Italy.* Paris: OECD.

———. 1994d. *Environmental Performance Reviews: Japan.* Paris: OECD.

———. 1994e. *Environmental Performance Reviews: United Kingdom.* Paris: OECD.

———. 1995a. *Environmental Performance Reviews: Canada.* Paris: OECD.

———. 1995b. *Environmental Performance Reviews: Netherlands.* Paris: OECD.

———. 1995c. *National Accounts, Main Aggregates, Vol. 1: 1960–1993.* Paris: OECD.

———. 1995d. *OECD Environmental Data: Compendium 1995.* Paris: OECD.

———. 1996. *Environmental Performance Reviews: United States.* Paris: OECD.

———. 1997. *Evaluating Economic Instruments for Environmental Policy.* Paris: OECD.

Porter, Michael E. 1990. *The Competitive Advantage of Nations.* New York: The Free Press.

Porter, Michael E. and Claas van der Linde. 1995. Green and Competitive: Ending the Stalemate. *Harvard Business Review* (September–October): 120.

Probst, Katherine, Don Fullerton, Robert E. Litan, and Paul Portney. 1995. *Footing the Bill for Superfund Cleanups: Who Pays and How?* Washington, D.C.: Brookings Institution and Resources for the Future.

Quarles, John. 1995. *Environmental Regulation: Brief for Reform.* Pfizer Corporation.

Repetto, Robert. 1995. *Jobs, Competitiveness, and Environmental Regulation: What Are the Real Issues?* Washington, D.C.: World Resources Institute.

Sorsa, Piritti. 1994. *Competitiveness and Environmental Standards.* World Bank Policy Research Working Paper 1249. Washington, D.C.: World Bank.

U.K. DOE (Department of the Environment and the Welsh Office). 1993. *Integrated Pollution Control: A Practical Guide.* London: HMSO.

U.S. CEQ (Council on Environmental Quality). 1996. *Environmental Quality: 25th Anniversary Report.* Washington, D.C.: Council on Environmental Quality.

U.S. DOE (Department of Energy). 1994. *International Energy Outlook 1994.* Energy Information Administration. Washington, D.C.: U.S. GPO, Table 5, 17.

U.S. EPA (Environmental Protection Agency). 1994. *National Air Pollutant Emission Trends, 1900–1993.* Office of Air Quality Planning and Standards. Washington, D.C.: U.S. EPA.

———. 1995. OSWER Directive 9355.7-04, 1. Office of Solid Waste and Emergency Response. Washington, D.C.: U.S. EPA.

WHO/UNEP. (World Health Organization and United Nations Environment Programme). 1992. *Urban Air Pollution in Megacities of the World.* Oxford: Blackwell Reference.

Weidner, Helmut. 1989. *A Survey of Clean Air Policy in Europe.* Berlin: Wissenschaftszentrum Berlin Fur Sozialforschung. FS II 89-301.

WRI (World Resources Institute). 1992. *World Resources 1992–1993: A Guide to the Global Environment, People and the Environment.* New York: Oxford University Press.

———. 1994. *World Resources 1994–1995: A Guide to the Global Environment, People and the Environment.* New York: Oxford University Press.

Yosie, Terry F. n.d. *The U.S. Environmental Management System: How Appropriate for Mexico?* Washington D.C.: E. Bruce Harrison Company.

10

Ability to Meet Future Problems

How effectively will the existing regulatory framework protect our environment in the next fifteen to twenty years? Specifically, what will be the condition of the environment assuming current regulatory levels and activities? Are there any foreseeable issues that are not being addressed by the current pollution control system? By examining available future projections, we hope to evaluate how well the pollution control system is poised to meet the challenges of tomorrow.

While many problematic events will be unforeseen, the pollution control system may be able to adjust and mitigate the expected environmental impacts. An added benefit of forecasting is that we may learn how improved information will guide the pollution control system in different directions. For example, if nonpoint sources are likely to be the primary cause of water pollution, we can make changes to strengthen our ability to control these sources. On the other hand, if point sources are the largest concern, the pollution control system might be modified in a different way. With these objectives, this portion of the study will examine several different issues that have been the traditional focus of environmental policy.

Future environmental issues can be divided into two general categories, those that can be predicted on the basis of current trends and those that will be new items on the environmental agenda. A case involving the former is relatively easy to understand: if we keep putting a known toxic into the environment, a negative environmental consequence is likely to result. New items, however, are often unpredictable. For example, the potential effects of chlorofluorocarbons (CFCs) on the ozone layer were not recognized until long after CFCs had developed key roles in the economy. It is unlikely that any amount of forecasting will eliminate environmental surprises. How the system is able to handle these "projected surprises" will be a determining factor in the efficiency and effectiveness of the pollution control system. Thus, one implication may be that flexibility and the ability to identify environmental issues as early as possible are desirable traits for the system.

While recognizing that some problems will be impossible to foresee, it is important to realize that to some extent forecasting can change unanticipated problems into predicted ones. If an issue is recognized before it becomes a problem, remedial action may be taken to avoid unwanted consequences. The more lead-time, the higher the chances of efficient and effective action. If forecasts are poor or nonexistent, we will be unable to anticipate future environmental challenges. Thus, the amount and quality of forecasting may be another criterion by which to assess the pollution control system.

Forecasting has a dual role in our evaluation. In order to evaluate the system, we must anticipate future environmental challenges. For example, if the current system allows auto emissions to grow beyond acceptable levels then, to that extent, the system is a failure. But, forecasting capability is itself also one of the criteria for judging the system.

Most of the projections examined in this chapter are aggregated at the national level. While this approach allows for a broad-based analysis, it also masks localized differences in environmental conditions, which are often significant.

FORECASTING AND UNCERTAINTY

Before embarking on our analysis, we should recognize some of the limitations of forecasting and how they impact its value. In the next few paragraphs, we investigate limitations common to all the analyses examined and discuss the importance of forecasting to achieving organizational goals.

Most of the few environmental projections that exist try to predict trends of perceived stressors (such as emissions to air or water, amount of waste disposed) rather than actual environmental quality. While in the broad sense trends in environmental stressors will determine trends in environmental quality, the relationship is not always obvious or direct. This is especially evident at the local level where variations in weather patterns, geology, and geography can influence the environmental impact of emissions. Thus, it must be recognized that all the projections examined below share the implicit assumption that higher levels of anthropogenic stressors are correlated with poorer environmental quality.

In creating projections of the future, it is necessary to understand where the current system is moving. This requirement implies knowledge of the current situation and its history, which together define the base from which the system evolves. In addition, one must identify the drivers of the system, and the forces that modify them. There are at least five major drivers of environmental stressors: population, economy, technology, policy, and social values. Although listed separately, these concepts are actually interconnected. For example, the size of the economy is affected by the number of people in a nation, and the structure of the economy is in part dependent upon the technology available and in use. Social values can shape policy, and policy defines the context for the economy and technological development.

The projections examined below make some assumptions regarding the system drivers. For the most part, the assumptions are implicit. Conditions were assumed either to be the same as today's or to continue along current trendlines. The only drivers that were consistently (or quantifiably) embodied in the projections are population and economic activity. Due to the seemingly unpredictable nature of policy, technology, and social value changes, these were considered to be the same as today's for most projections.

The dynamic nature of these drivers and their inherent uncertainties mean that any prediction of future trends is likely to be wrong. A brief examination of the factors influencing air emission projections illustrates this point. Projections for *EPA National Air Pollutant Emission Trends, 1900–1994* (U.S. EPA 1995b) are constructed using a deceptively simple equation:

$$\text{(Level of activity)} \times \text{(Emissions factor)} \times \text{(Amount of control)} = \text{(Total Emissions)}$$

Economic and demographic projections, each with its own level of uncertainty, are used to calculate activity level. Standard emissions factor calculations are used by EPA, but each type of combustion technology has its own factor. Thus, there must be some way of projecting how the distribution of different types of combustion technology will change over time. The same is true for control technologies. In addition, the availability of technology and the nature of government policies will influence all of these factors simultaneously. Add to this the uncertainty of how emissions actually affect ambient air quality and the result is a very weak prediction of future environmental conditions.

While this uncertainty may seem to imply that there is little value in forecasting, the reverse is true. In the management literature several authors suggest that in the face of such uncertainties organizations must use more strategic planning to achieve their goals. Although making comparisons between firms and governments is dangerous, one can draw some useful parallels. Because of increasing uncertainties, the importance of strategic planning for an organization is essentially heightened. Among the approaches suggested are "higher emphasis on 'qualitative' analysis and scenario building, increasing the firm's information processing capacity, higher frequency of environmental scanning, the use of interactive management control systems and a high degree of openness in the planning process" (Karagozoglu 1993, 335). Used in this way, forecasting acts as a feedback mechanism to alert organizations to changing conditions and can guide adjustments to behavior.

The importance of projections, therefore, lies not in their accuracy, but in their ability to inform discussion about current directions and policy options. The National Acid Precipitation Assessment Program Interim Assessment eloquently summarizes the value of forecasting: "Emissions projections are best viewed as conditional statements about future expectations. If, for example, a certain growth rate is different or if different technologies are deployed, then the emissions estimates will be different. These types of conditional statements about the future can be valuable aids to structured thinking about future energy, economic, and environmental conditions; even more important, they can shed light on the possible long-term effects of alternative policy actions. To the extent that any particular projections are viewed as precise predictions of the future, however, they can become impediments to constructive analysis" (Placet and others 1987, 3-3). "The challenge faced by ecological economic modelers is to first apply the models to gain foresight and then respond to and manage the system feedbacks in a way that helps avoid any foreseen cliffs" (Costanza and others 1993, 551). With this understanding of forecasting, we proceed to examine future trends in various areas of the pollution control regulatory system.

AIR POLLUTION

Air pollution arises from a number of sources, both natural and anthropogenic (caused by humans). Thus air quality may not be directly related to the quantity of anthropogenic emissions. EPA estimates that annual biogenic volatile organic compound (VOC) emissions are greater than anthropogenic emissions (U.S. EPA 1995b, 9-1), implying that trends in air quality may not follow VOC emissions exactly, all other things being equal. Weather patterns and geography also can have an enormous impact on air quality, which highlights another concern: localized effects of emissions and transport of pollutants can lead to erroneous conclusions about trends in air quality. While some sources and types of pollution (such as SO_2 from tall stacks) can be distributed widely, others (such as auto emissions in the Los Angeles basin) tend to be more localized. It is difficult for national projections to take into account local effects (see Chapter 5).

Recognizing the limitations of the data, EPA publishes projected emissions trends for several criteria pollutants by source category as part of its *National Air Pollutant Emission Trends*. Forecasts are based on the assumption that no further regulation is enacted beyond what exists today. In addition, some technological change is embedded in the projections in that different source categories grow at different rates. The trend data are plotted in Figure 1. The CO_2 data in Figure 1 are from the Department of Energy (U.S. DOE 1995, 95), whose figures do not include nonenergy anthropogenic sources. (Since DOE estimates that more than 98 percent of total anthropogenic emissions of CO_2 are due to energy usage [U.S. DOE 1994b, 2], this is a reasonable approximation.)

Data for VOC, NO_x, and CO all show decreased emissions over the next five to seven years, followed by smaller increases through the year 2010. By the year 2000, CO emissions are expected to fall to 81 percent of their 1990 level value and then rise to 87 percent in 2010. VOC emissions trends are similar to CO, declining to 83 percent of their 1990 value by 2000 and then increasing slightly between 2000 and 2010 to 86 percent of their 1990 level. NO_x emissions are more stable, reaching a minimum of 89 percent of 1990 emissions in 2000. In 2010, they are expected to be 94 percent of their 1990 value.

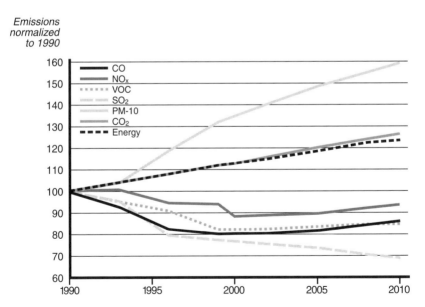

Figure 1. Normalized U.S. Total Anthropogenic Air Emissions Projections (1990 = 100).

Sources: U.S. DOE 1995; U.S. EPA 1995b.

Trends for PM-10 and CO_2 show increasing emissions over the entire projection period, while SO_2 emissions fall steadily over the same time frame. CO_2 and PM-10 emissions are expected to grow to 126 percent and 159 percent of their 1990 levels by the year 2010, respectively. SO_2 emissions, on the other hand, are predicted to fall to 70 percent during the same period.

As pointed out earlier, natural emissions of some pollutants are substantial. Table 1 estimates the share of total emissions attributable to anthropogenic sources for each of the pollutants summarized. Data on emissions due to energy use (a subset of anthropogenic sources) are also given. These percentages are intended to give the reader a sense of magnitude.

From Table 1, it is apparent that natural sources of CO_2 and VOCs are greater than anthropogenic sources. As mentioned, EPA estimates that biogenic sources may account for slightly over half of the national VOC emissions (U.S. EPA 1995b, 9-1). CO_2 emissions from manmade sources are even a smaller proportion of the annual quantity released. Estimates suggest anthropogenic emissions account for between 3 and 7 percent of CO_2 released annually (U.S. DOE 1993, 4). This implies that reducing emissions from anthropogenic sources will not necessarily eliminate air quality concerns due to these pollutants. However, in the case of CO_2 emissions, the concern is that natural sources of CO_2 emissions were historically balanced by natural sinks, and that now manmade sources are leading to an increase in atmospheric CO_2 (U.S. DOE 1994b, 2).

Table 1. Estimated Contribution of Anthropogenic Sources to Air Pollutant Emissions (as Estimated Percentages).

Pollutant	Total emissions due to anthropogenic sources	Anthropogenic emissions due to energy usage	Total emissions due to energy usage
CO	99+[a]	80–85[a]	80–85[b]
NO_x	85–95[c]	90–95[a]	80–85[c]
VOC	40–50[d]	35–45[a]	15–25[a]
PM–10	90–98+[a]	3–5[a]	3–5[a]
SO_2	90–98+[e]	85–95[a]	80–90[d]
CO_2	3–7[f]	98+[b]	3–7[f]

[a]Calculated from U.S. EPA 1994b. Wildfires are counted as anthropogenic sources.

[b]Calculated from U.S. DOE 1994b and U.S. EPA 1994b.

[c]Calculated from U.S. EPA 1994b and Placet and others 1987. NAPAP data from the latter source, relating anthropogenic emissions to natural sources, are for all North America.

[d]Calculated from U.S. EPA 1994b and Placet and others 1987.

[e]From Placet and others 1987.

[f]Calculated from U.S. DOE 1994b and U.S. EPA 1994b. Data relating anthropogenic emissions to natural sources are for the world.

Table 1 also indicates that energy is the predominant source of anthropogenic emissions for most criteria pollutants and CO_2. (Data for CO_2 emissions due to fuel consumption are compiled by the Department of Energy as part of their *Annual Energy Outlook*.) PM-10 emissions are the exception. In 1990, EPA estimated that energy use accounted for less than 5 percent of total anthropogenic PM-10 emissions. Nearly 75 percent of anthropogenic emissions for that year are attributable to fugitive dust from roads and other sources (U.S. EPA 1995a, 3-15). For the remainder of the pollutants listed, energy use accounts for between 35 and 98 percent of anthropogenic emissions. Thus, anthropogenic air emissions trends will be significantly influenced by trends in energy consumption.

Although energy consumption is the dominant cause of air emissions, specific sources of anthropogenic emissions vary by pollutant. This implies that several different strategies may be necessary to achieve reductions in all the pollutants considered. While focusing on stationary sources may be the most effective way of reducing SO_2 emissions, the same approach will not likely be successful in achieving significant reductions in CO emissions.

Typically, air emissions sources are divided into two categories, mobile and stationary. All mobile source emissions are due to energy consumption, while stationary source emissions are caused either by energy consumption or other activities. Despite somewhat substantial changes in overall emissions, EPA anticipates that the relative shares of each source will remain fairly consistent between 1990 and 2010.

Energy Use and Air Pollution

This analysis highlights the contribution of energy in general, emphasizing the different contributions of mobile energy sources (primarily highway vehicles, boats, and airplanes) and stationary energy sources (such as industrial users and commercial and residential heating). For VOC and CO emissions, mobile sources contribute the bulk of emissions due to energy consumption, while stationary energy sources account for most of the SO_2 emissions. Energy-based NO_x emissions are split evenly between stationary and mobile sources; for PM-10, although energy accounts for a small percentage of the total emissions, stationary exceed mobile sources by a factor of three.

Not surprisingly, most of the emissions trends due to energy usage shown in Figure 2 are similar to the total anthropogenic emissions trends in Figure 1. The exception is PM-10, whose emissions trend follows that of NO_x, VOC, and CO. These four pollutants show a decrease in emissions over the next five to seven years, followed by a smaller increase through the year 2010. As with total anthropogenic emissions, energy-based SO_2

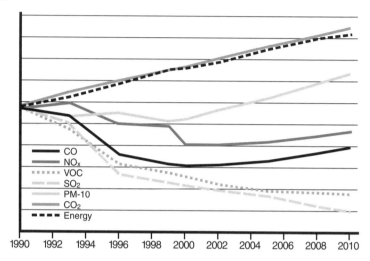

Figure 2. Normalized U.S. Energy Demand and Emissions Due to Energy
(1990 = 100).

Sources: U.S. DOE 1995: U.S. EPA 1995b.

emissions fall for the entire projection period, and CO_2 emissions continually increase.

Overall, these trends indicate substantial reductions in pollution per
BTU of primary energy demand. Despite a 24 percent increase over 1990
levels in energy use, the projected emissions of CO, VOC, SO_2, NO_x, and
PM-10 in the year 2010 are lower than in 1990. CO_2 emissions increase
over this time period but at a lower rate than primary energy consumption. However, beyond 2010, without changes to the current projected
energy and efficiency trajectories, higher consumption will likely overcome the gains that have been made.

Criteria air pollutants are not the only types of air emissions. Hazardous air pollutants and air toxics (as defined by the emissions of chemicals on the Toxics Release Inventory list) are two overlapping categories
that complete the picture of air quality. *National Air Pollutant Emission
Trends, 1900–1994* provides 1990 baseline inventories for thirteen hazardous air pollutants by source categories. While projections for these and
other air pollutants are not currently available, the 1990 inventories do
provide some insight. First of all, the contribution of energy usage to
overall emissions ranges from nearly all (alkylated lead compounds) to
essentially none (hexachlorobenzene). Thus, the effectiveness of regula-

tion aimed at energy usage only will vary by particular air toxic. Secondly, Section 112 (k) of the 1990 Clean Air Act Amendments requires EPA to identify " area sources accounting for at least 90 percent of total emissions of at least thirty hazardous air pollutants that present the greatest threat to urban populations" (U.S. EPA 1995b, 10-3). The focus on area sources affecting urban populations implies that mobile sources (and by extension, energy usage) will play a potentially significant role in achieving this requirement.

In conclusion, it is apparent that energy use plays a significant role in air pollution, especially in terms of criteria air pollutants. Energy usage must be addressed to make any substantial progress in reducing emissions of these pollutants. If, as projected, energy demand grows nearly 25 percent in the next twenty years, substantial reductions in emissions per BTU are necessary to hold pollution rates constant, let alone reduce them. The good news is that this reduction is what is expected. The bad news is that beyond 2010, the gains in emissions per BTU are expected to be overwhelmed by increased energy demand. While any number of technological innovations may reduce pollution per BTU beyond that projected, the most obvious way to reduce emissions is to reduce energy demand. A reduction in energy demand through increased efficiency can have a profound impact on air emissions.

Energy is not the only important factor in air pollution. The particular regulatory approach will depend on the pollutant to be controlled. Whereas historical command-and-control regulation can be effective in addressing SO_2 emissions from a relatively small number of electric utilities, this method will be less able to handle numerous, dispersed automobiles each contributing a small portion of the total CO emissions. To address this sort of pollution may require a change in individual behavior, which in turn implies a different policy approach.

WATER POLLUTION

Water is perhaps the most renewable resource on earth. Its global amount is essentially constant, regardless of human activity, and it is constantly moving from the atmosphere to oceans and land (through precipitation) and back (through evaporation and transpiration). Evaporation and transpiration act as purification processes, leaving salt and other impurities behind as water is vaporized. Although water can be stored in plants and animals, it is never truly removed from the system since it is eventually released through waste elimination or decomposition. Two-thirds of the precipitation that falls on the conterminous United States is transferred directly back to the atmosphere through evaporation or transpiration.

The remainder eventually returns to the oceans, either as runoff into surface water (rivers, lakes, and so forth) or by flowing through the soil into underground aquifers (Frederick 1995, 14–15).

The quality of fresh water, while often addressed as a separate issue, is an important parameter for understanding our ability to meet resource demand. If surface or groundwater sources are polluted, they may be unusable, reducing available freshwater supplies. However, the relationship between quality and quantity is not unidirectional; high demand can impact quality. Elevated consumption of surface water can impair instream uses such as navigation, aquatic life and recreation, while excess use of water from underground aquifers can lead to degradation of supply due to salt water intrusion as is the case in a number of coastal regions (Frederick 1995, 19). The link between quality and quantity is further evidenced by the slowing of demand for water driven by policy changes aimed at improving water quality. For example, being charged "the full cost of treating waste water sent to municipal treatment facilities" as part of the Clean Water Act, seems to lead "industrial users [to] adopt internal recycling strategies to reduce their waste flows and thus municipal waste treatment fees" (Guldin 1989, 42). Water quality and quantity are interdependent; changes in one will affect the other.

Generally speaking, fresh water for human use comes either from surface water or groundwater sources.[1] Groundwater aquifers are either confined or unconfined: unconfined aquifers are continually replenished by water flows through the earth's crust while confined aquifers have essentially no replenishing flows. If the rate of withdrawal from an unconfined aquifer exceeds the rate of replenishment, the level of the aquifer will fall, resulting in less groundwater inventory (water in reservoirs) and higher pumping costs to access the remaining water. Withdrawals from confined aquifers result in a net loss of groundwater inventory, since by definition the aquifers are not replenished. Thus, as the rate of extraction from unconfined aquifers exceeds the rate of replenishment (or as confined aquifers are mined), available groundwater supplies are reduced.

Surface water sources present a different set of supply challenges. The instream uses of water, such as navigation, hydropower, commercial fishing and recreation, require a certain minimum amount of water. Below that minimum, lakes and rivers can become impassable, and ecosystems can be damaged. Thus, instream flow and inventory requirements set the minimum threshold required for adequate instream uses. When considering water supply issues, these thresholds are part of the "demand" for water.

The offstream demand for water can be divided into consumptive and nonconsumptive uses. Consumption refers to water that is withdrawn from a source and not returned in a usable form. Thus, the portion of water used for irrigation that evaporates or is absorbed by crops is considered to be consumed. Although it is not removed from the hydrologic process, such water is not usable again until it falls as precipitation. The importance of consumption to water issues is obvious. Nonconsumptive demands for water can be satisfied repeatedly with the same parcel of water, whereas consumptive uses remove water from the watershed. Thus, from a supply standpoint, the most appropriate measure of water demand is not withdrawals but consumption. However, overall water withdrawals are important in that nonconsumptive withdrawals are rarely returned in exactly the same condition as they are removed.

We next examine future trends in water supply/demand relationships and quality with the purpose of discovering what they tell us about the pollution control system and how it is poised to address future challenges. While some of the discussion will be based on nationally aggregated data, water resource issues have an important regional component. Unless transfer facilities are available, rain that falls in one watershed will be useless to another. In addition, seasonal and annual variations in precipitation and snow melt affect the available supply at any given time. Water supply problems arise from vagaries in spatial and temporal distribution. The flooding in the Midwest in 1996 and 1997 is an extreme example of oversupply, while the drought in California between 1987 and 1993 is an example of undersupply. From a quality perspective, the issues can be even more localized as the contamination of groundwater is often an isolated event. Thus, a focus on national averages of precipitation and quality does not fully represent the true character of water supply and quality issues.

Supply and Demand

The only recent government projection of water supply/demand in the United States was undertaken as part of the *1989 Forest Service RPA Assessment* (Guldin 1989). In this analysis, actions to expand supply through conventional methods (such as dams and reservoirs) or other means (cloud seeding, desalination, recycling) were assumed to have negligible impact. The report includes projections for water demand through the year 2040 (broken down by water resource region, use, and source), which were developed using regression analysis of data between 1960 and 1987. Since this type of analysis projects future conditions by extending past trends, any changes that result in a major shift in water use (changes in water rights or allocation methods, new technologies) are not included. National water demands by use based on the Forest Service assessment are summarized in Figures 3 and 4.

Trends show both freshwater withdrawals and consumption increasing to about 40 percent above 1980 levels by 2040, with the ratio of consumption to withdrawals throughout the sixty-year period hovering near 30 percent. From the standpoint of withdrawals, thermoelectric power generation is the largest user for the projected time period, followed closely by irrigation. Irrigation, however, represents by far the largest consumptive use of water resources. Although its relative importance falls throughout the projection period, by 2040 irrigation is still projected to account for 70 percent of consumptive water use in the United States. The relationship between water and agriculture will be examined in more detail below.

In addition to demand projections, the assessment examines supplies in concert with demand to identify potential water surpluses and deficits. Two different supply conditions (average rainfall years and below average rainfall years) are matched with two different demand conditions (instream flow required for optimal fish and wildlife habitat conditions and instream flow required for good survival conditions for fish and wildlife habitat, each coupled with consumption projections). (*Instream flow* represents water in rivers and streams flowing out of the watershed. *Renewable supply* is the precipitation that reaches aquifers or that runs off into surface water supplies. *Groundwater overdraft* represents the quantity of water withdrawn from aquifers in excess of the recharge volume.) Mathematically,

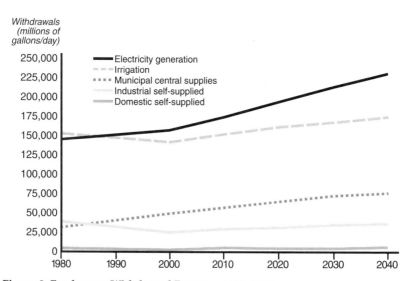

Figure 3. Freshwater Withdrawal Patterns, 1980–2040.

Note: Withdrawals for livestock watering are less than 1 percent and are not shown.

Source: Guldin 1989.

$$\text{Surplus or Deficit} = \text{Supply} - \text{Demand}$$

where

$$\text{Demand} = \text{Consumption} + \text{Instream Flow Requirement} \\ + \text{Reservoir Evaporation}$$

and

$$\text{Supply} = \text{Renewable Supply} + \text{Groundwater Overdraft} + \text{Net Imports}$$

A deficit indicates that demand is expected to exceed projected supply. When a deficit is shown, one of the initial assumptions used in creating the balances is necessarily violated (for example, instream flow falls below the "required instream flow," groundwater draft is higher than predicted, or stored inventories are depleted).

For the conterminous United States, the data show enough precipitation so that supply exceeds projected demand for the entire forecast period. However, six out of eighteen regional watersheds—Lower Colorado, Upper Colorado, Rio Grande, Great Basin, California, and Lower Mississippi—are expected to experience shortages at some time over the forecast period. For the affected watersheds, the magnitude of projected shortages grows with time due to increasing demand for water.

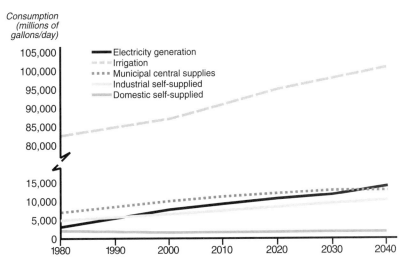

Figure 4. Freshwater Consumption Patterns, 1980–2040.

Note: Consumption for livestock watering is 2 percent or less and is not shown.

Source: Guldin 1989.

Table 2 presents total demand (including instream flow required for "good habitat survival") as a percentage of total supply during below-average-rainfall (dry) years for all basins expected to show deficits. (*Average year* data were excluded from Tables 2, 3, and 4 for the sake of brevity.) The measure is intended to give a reader a sense of the magnitude of the deficits. The values in italics represent conditions where the sum of consumption and evaporation exceeds total supply. At first glance, this outcome seems impossible (how can you use more than you have), but physically, this situation can arise when reservoirs are drawn down, thus allowing the sum of consumption and evaporation to temporarily exceed "total" supply (the sum of renewable supplies, groundwater overdraft, and net imports). Obviously, draining reservoirs is a limited and risky strategy for managing water resources.

From Table 2, it is apparent that the Lower Colorado and Rio Grande regions will have the most significant dry year shortages. For the Lower Colorado, enormous and increasing deficits are projected for years with either average or below average precipitation. As shown in Table 2, water demand in that region is more than twice the projected supply during dry years, implying major shortages. Similarly, the Rio Grande region is expected to experience significant water shortages in both average and dry years. There, dry year demand exceeds supply by 25 to 37 percent over the forecast period. Table 3 presents total consumption as percentage of total supply, in an attempt to understand the magnitude of human demands on water resources. Not surprisingly, offstream consumptive uses will take 70 percent or more of available water in the Lower Colorado and Rio Grande basins during dry years. Thus, the effects of shortages will be felt by the nearly 7 million people who currently live in the Lower Colorado and Rio Grande water resource regions. One probable result of the imbalance is that withdrawals from the river basin will cause

Table 2. Total Water Demand As Percentage of Total Supply: Dry Years.

Basin	Total demand as percentage of total supply				
	2000	2010	2020	2030	2040
Lower Colorado	217	225	226	229	232
Rio Grande	125	129	132	135	137
Upper Colorado	N/A	N/A	101	104	106
Great Basin	106	109	113	115	117
California	114	118	121	124	126
Lower Mississippi	103	104	105	106	108

Notes: N/A means no deficit projected. The values for the Lower Colorado basin represent conditions where consumptive and evaporative demands exceed total supply.

Source: Guldin 1989.

instream flows to fall below those necessary for good instream habitat survival. In addition, groundwater overdraft in the Lower Colorado will continue to increase, leading to even faster resource depletion (the aquifers in the Rio Grande basin are not capable of significant increases in withdrawals, and thus are not expected to experience significant over-draft). However, neither of these events is likely to be sufficient to eliminate the resource imbalance. Reducing consumption, especially from irrigation, is likely to be the only way to bring the regions' water demands into balance with available supply (Guldin 1989, 81–82).

In 1985, consumption due to irrigation was 87 percent and 88 percent of total consumption for the Lower Colorado and Rio Grande basins, respectively. The national average for that year was 78 percent. As irrigation is both the largest consumer and the lowest value use of water, it likely will provide the most significant reductions in consumption. For the Rio Grande region, a 16 percent reduction in irrigation consumption from 1985 to 2000 will eliminate projected deficits to 2040 with average rainfall and reduce the dry year deficits to manageable levels. For the Lower Colorado, the situation is more severe: a similar reduction in irrigation would not be enough to eliminate deficits. There, the projected deficit for most years nearly equals consumption, implying that eliminating *all* consumption would barely provide the river with adequate flow for optimal (or good survival of) fish and wildlife habitat for average (or dry) years. To bring demand into balance with renewable supplies will require a reduction in all types of consumption, in addition to other steps. As the greatest consumptive use for this region, irrigation must be addressed to achieve this goal (Guldin 1989, 81–83).

Table 4 highlights the key role agricultural consumption (in particular, irrigation) plays in water supply issues by showing deficits as a percentage of agricultural consumption. The numbers represent how much agri-

Table 3. Projected Water Consumption as Percentage of Total Supply: Dry Years.

| Basin | *Total consumption as percentage of total supply* | | | | |
	2000	*2010*	*2020*	*2030*	*2040*
Lower Colorado	121	131	132	136	140
Rio Grande	70	74	77	80	83
Upper Colorado	N/A	N/A	40	42	44
Great Basin	68	71	74	77	79
California	58	61	64	67	69
Lower Mississippi	13	14	15	16	17

Notes: N/A means no deficit projected. The values for the Lower Colorado basin represent conditions where consumptive and evaporative demands exceed total supply

Source: Guldin 1989.

cultural demand would have to be reduced to eliminate the water deficits, all else being equal. (In practice, it is unlikely that reduction in agricultural demand will be the only step taken to eliminate deficits.) Here again, the Lower Colorado basin stands out, with deficits greater than agricultural demand. Reducing agricultural consumption alone is insufficient to eliminate deficits.

The deficits for the Upper Colorado, California, Great Basin, and Lower Mississippi regions are much less severe, as shown in Table 2. Only in California and the Great Basin do total demands exceed total supply by more than 10 percent at any time during the forecast period. For this reason, relatively minor adjustments in consumption growth will eliminate projected shortages. As with the Lower Colorado and Rio Grande, reduced or more-efficient irrigation will likely play a significant role in reducing overall consumptive use. In these regions, irrigation ranges from 72 percent to 92 percent of all consumption; as in other areas, irrigation is the least valued use of water. One interesting point is the existence of deficits in the apparently water-rich region of the Lower Mississippi, which highlights the growing pressure on water resources (Guldin 1989, 84).

Water Quality

As indicated in Chapter 5, existing national water quality statistics are insufficient to show generalized trends. However, the existing data clearly indicate that, contrary to popular conception, the major source of impairment to water resources is not industrial discharge but nonpoint source pollution, particularly agricultural practices. For example, nitrate and phosphate concentrations in stream runoff have been statistically associated with various agricultural activities including fertilizer use (Smith, Alexander, and Wolman 1987, 1612–13).

Table 4. Projected Water Deficits as Percentage of Agricultural Consumption: Dry Years.

Basin	Projected deficits as percentage of agricultural consumption				
	2000	2010	2020	2030	2040
Lower Colorado	112	115	113	113	113
Rio Grande	40	44	47	50	52
Upper Colorado	N/A	N/A	3	9	15
Great Basin	9	14	18	21	24
California	27	31	36	39	42
Lower Mississippi	19	28	35	61	67

Notes: N/A means no deficit projected. The values for the Lower Colorado basin represent conditions where consumptive and evaporative demands exceed total supply
Source: Guldin 1989.

Similarly, agricultural activities have been identified as a significant threat to groundwater by both the EPA (U.S. EPA 1994b, 72) and USGS (1988, 136–41). As an example, one study found a statistically significant relationship between fertilizer use and nitrate in groundwater (Chen and Druliner 1988, 103–8). With more than 50 percent of the population currently relying on groundwater for domestic uses (Solley, Pierce, and Perlman 1993, 26), management of pollution sources that affect groundwater quality is an important and growing issue. Point source pollution will remain a concern in specific localities, but most observers agree that nonpoint sources will be the primary contributors to both surface and groundwater pollution for the foreseeable future. In particular, agricultural practices will continue to play a key role in water quality.

The major agricultural impacts on water quality are a function of the amount of chemicals used in production (pesticides and fertilizers); the amount of land used in production; the amount (and intensity) of irrigation; and the rate of soil erosion. While seemingly independent, these parameters are actually interdependent. For example, the amount of chemicals used in production and the rate of soil erosion are strongly influenced by the amount of land cultivated.

The two primary determinants of agricultural practices are agricultural policy as spelled out in the farm bill (enacted every five years) and the international market for agricultural products. Together these two factors affect the amount of land used in agriculture, the amount of agricultural chemicals used, and the amount of irrigation used in crop production.

Perhaps the most direct impact of these factors is their effect on the amount of land in crop production. Through programs such as the Conservation Reserve Program and Acreage Reduction Program, farmland is taken out of production. Similarly, increased competition in international agricultural markets will tend to drive prices down, leading to increased amounts of acreage allowed to lay fallow. While export totals are impossible to predict from year to year due to chaotic variables like the weather, in general increasing crop production in Russia and Asia will likely dampen the U.S. export market. Regardless of the cause, a reduced amount of land used in crop production will tend to reduce the overall impact of agriculture on water quality, all other things being equal.

Farm bill activities aimed at curtailing agricultural chemical usage (such as integrated pest management) will also have an effect on water quality. To the extent that these programs are successful in reducing the amount of chemicals applied, they will mitigate the environmental effects of agriculture. Other national agricultural policies affecting the amount of land irrigated have a similar impact.

Agricultural activities that impact water quality will change, then, in response to changing agricultural policies and international markets.

Thus, future water quality will depend in part upon agricultural policy, an area traditionally outside the purview of EPA and its state counterparts. Furthermore, international markets for agricultural products generally are beyond the control of the national government, imposing an even greater challenge to efforts to mitigate agriculture's effect on water quality. Any attempt at mitigation will have to consider these issues.

In summary, existing water quality data show that while point source pollution will continue to need attention, the primary challenge for the pollution control system now and in the future is to develop a system that addresses the numerous, geographically dispersed sources that impair water quality. In particular, agriculture will continue to play a major role. The primary drivers of agriculture are the policies contained in the Farm Bill and international markets for agricultural products. These two factors are both somewhat unpredictable and beyond the scope of the traditional "pollution control system."

Just as water supply issues are first and foremost regional issues, the same is true of water quality. Great Lakes pollution illustrates this point. As reported in *National Water Quality Inventory 1992* (U.S. EPA 1994b), atmospheric deposition is the primary source of water quality impairment for the Great Lakes, but it is not identified by EPA as one of the top five causes of pollution of other lakes and reservoirs. In fact, comparing the five leading sources of impairment for the Great Lakes and other lakes and reservoirs, we find only one entry (urban runoff/storm sewers) on both lists. Because threats to water quality will vary depending on the water body, a variety of approaches will likely be necessary to improve water quality significantly, implying a flexible pollution control system.

Water Quality and Quantity Summary

Taking quantity and quality data together, a picture begins to emerge. Water resources in the United States are theoretically sufficient to meet all consumptive needs, on an *average* basis. For the conterminous United States , the annual amount of precipitation that reaches our surface and groundwater supplies is enough to supply all the water necessary for the foreseeable future. The threats to water resources arise from the manner in which water is allocated between different uses, the geographic differences in supply and demand, and the damages to existing supply due to pollution. The current institutional framework for allocation has been criticized for presenting barriers to efficient use and protection of those resources (Frederick 1992, 25). Additionally, the traditional end-of-pipe focus on pollution is becoming less effective at eliminating the major threats to water quality that come increasingly from nonpoint sources. The EPA's historical focus on point sources, consistent with its statutory authority, only addresses part of the problem. As we have shown, the ties between quality and quantity imply that these issues are not independent.

Land use has been of primary importance to water issues in the past and will continue to be critical in the foreseeable future. Agriculture in particular has major impacts on both the amount of water consumed and the pollution of water resources. These factors point toward a role for an integrated approach to water supply and quality issues that emphasizes ties to land use, particularly agricultural land use. EPA and others, recognizing these relationships, have begun to address this situation through interagency efforts to improve land management practices. The importance of such efforts to achieving overall water quality and allocation goals will grow as nonpoint sources of pollution become increasingly significant and as increasing demand encroaches upon limited supplies. The key role of the USDA in determining agricultural practices is particularly important. The future of water pollution control is a function of activities that are not presently within the traditional pollution control system, either because they are outside EPA's jurisdiction (such as agriculture) or because neither EPA nor Congress has focused on them (such as air deposition and nonpoint sources).

MUNICIPAL SOLID WASTE

As explained in Chapter 5, EPA defines municipal solid waste (MSW) as a subset of nonhazardous solid waste. It comprises the waste handled by local and state governments, excluding construction, and demolition waste, industrial process waste, automobile bodies and combustion ash. In the context of MSW, the term *generation* refers to the amount of material that enters the waste stream; it excludes such things as yard waste placed in backyard compost heaps, which is considered one means of source reduction. Of the waste that is generated, some gets recycled. The remainder of the waste generated is then disposed by either incineration or landfilling. Figure 5 represents the cycle schematically. As part of the *Characterization of Municipal Solid Waste in the United States* (U.S. EPA 1996), the EPA includes projections of these MSW quantities through the year 2010. These data, shown in Figure 5, were derived by extending past MSW trends and adjusting for expected future conditions.

MSW is an issue for several reasons. Eventual disposal of net waste (that is, the amount of waste generated that is not recycled) is currently limited to two options, each with its own set of environmental concerns. The first option is land disposal in a sanitary landfill, which requires significant space. Although adequate landfill capacity was a major concern in the late 1980s (and helped focus attention on MSW management),

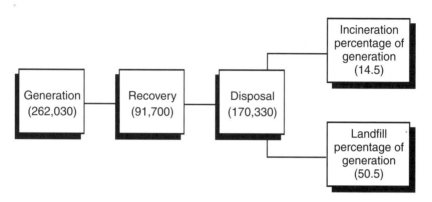

Figure 5. Muncipal Solid Waste System Flow, 2010.

Note: Data for 2000 assume 30 percent recovery and 25 percent reduction (from 1994) in yard trimmings entering the muncipal solid waste (MSW) stream. Data for 2010 assume 35 percent recovery and 9 percent increase (from 2000) in yard trimmings entering MSW stream. All assumptions are midrange estimates.

Source: U.S. EPA 1996.

recent studies by the National Solid Waste Management Association show that most states estimate they have more than ten years of capacity in existing landfills (Repa and Blakely 1996, 4–5). The most significant environmental concern with landfill disposal is leaching of contaminants (such as organic compounds, nutrients, and toxics) into surface or groundwater. Active landfills have been identified as a threat to groundwater quality in all states (USGS 1988, 136–42). To help address leaching concerns, EPA issued regulations in 1991 that set minimum standards for MSW landfills (40 CFR Section 258). Other environmental concerns include release of methane from landfills, and the function of the site as a food source for disease vectors such as rats and flies.

The second option for MSW disposal is incineration of waste to reduce waste volume, with the ash residue then landfilled. The primary environmental issues with this method are incinerator stack emissions and incinerator ash, both of which may contain heavy metals, dioxins, and other toxics that can enter the ecosystem if not properly managed. A recent ruling by the Supreme Court required EPA to regulate incinerator ash as a hazardous waste if it meets the criteria for toxicity (*City of Chicago, et al., v. Environmental Defense Fund, et al.* 114 S. Ct. 1588 [1994]). Previously, such ash was exempted by EPA from regulation as a hazardous waste.

In addition to the immediate environmental concerns with its disposal, trends in MSW generation and disposal can provide a perspective on the use of materials within a nation. While not a comprehensive view, as it neglects the use of most materials in the production sector of the economy, the trends do afford a glimpse of consumption patterns. In developing the projections, EPA used two time periods with different assumptions associated with each. During the first period, from 1994 to 2000, it is assumed that recovery of MSW for recycling and composting is a constant 30 percent of the total amount generated, the amount of yard trimmings generated falls 25 percent, and the generation of other materials (glass, paper, and metals) follows existing trendlines. Using these assumptions, per capita generation levels off at about 4.4 lbs./person/day over the time period while the absolute amount of MSW generated increases. In contrast, the disposal of MSW on both a per capita and absolute basis falls over the same six-year time frame, due to increased rates of recovery.

By 2000, EPA assumes that all regulatory incentives (at all levels of government) are in place to eliminate yard trimmings from the MSW stream. Thus, between 2000 and 2010, yard trimmings entering the MSW stream increase in proportion to population. Additionally, the recovery rate is expected to climb to 35 percent (that is, recovery continues to grow but at a lower rate than the 1994 to 2000 period). The result is an increase in per capita generation to 4.8 lbs./day, which leads to an increase in the quantity of MSW generated. The per capita disposal rate for this period increases very slightly, but remains below the 1994 rate of 3.4 lbs./person/day. However, due to expected population growth, the quantity of waste disposed in 2010 is greater than 1994.

As discussed in Chapter 5, the per capita generation of MSW increased from 2.7 lbs./day in 1960 to 4.4 lbs./day in 1994. The slower growth in per capita generation over the entire forecast period represents a change in MSW trends. The shift is even more striking in projections of net disposal. Between 1994 and 2000, increased recycling and municipal composting combined with lower generation rates lead to an expected decrease in the absolute quantity of waste disposed. While this absolute reduction is fleeting (the amount of waste disposed in 2010 is anticipated to exceed that of 1994), the per capita disposal rate is expected to remain at 3.1 lbs./day, down from 3.4 lbs./day in 1994.

Both generation and disposal trends illustrate the impact recycling and other waste management programs are having on the MSW system. Despite projected increases in economic activity, per capita generation growth slows and disposal rates actually decline. These changes imply fewer environmental impacts and use of fewer resources by U.S. consumers than if historical rates of generation and disposal prevail over the forecast period. Note that this situation does not necessarily mean the material efficiency (the amount of materials used per dollar of GDP) of the economy as a whole has improved. While there is evidence of a declining demand per dollar of GDP for some industrially important minerals, data on the overall dematerialization of the economy are ambiguous (Ausubel 1996, 174; Bernardini and Galli 1993, 445).

The distribution of disposal methods is expected to remain fairly constant, while the absolute quantity of waste incinerated will increase about 17 percent between 1994 and 2010. This increase is based on the assumption that no new incineration capacity is added during the period and the extension of current trends in the combustion of certain source separated waste material. Note that as a percentage of generation, both incineration and landfill disposal fall over the time period, reflecting the increased amount of recycling and recovery.

In summary, MSW projections show that increased attention to waste management systems and the recycling boom of the late 1980s and early 1990s is expected to lead to slower growth of per capita generation and a decline in per capita disposal. These trends reflect reduced materials consumption by the consumer sector of our economy.

HAZARDOUS WASTE

As discussed in Chapter 5, existing hazardous waste generation and disposal data are inadequate to develop trends. However, indications that the amount of waste generated may be decreasing are evidenced by EPA Biennial Reporting System data and anecdotal industry observations (see Chapter 5). Future projections of hazardous waste data do not exist. As part of the EPA's Capacity Assessment Plan (CAP), the agency estimates commercial hazardous waste management capacity in relation to demand on these facilities; however, it does not attempt to project changes in generation per se. In fact, EPA states that "[g]enerally, based on Agency recommendations, States reported in their CAPs that demand on subtitle C management capacity ... remained constant between 1993 and 1999" (EPA 1994a, 7). While this does not necessarily mean generation rates are expected to be constant (generators could handle more waste on-site), it also does not provide a basis for estimating future quantities of hazardous waste. In any case, using essentially the same demand as 1993 (the base year), the report shows sufficient capacity in our existing commercial hazardous waste management system through the year 2013. In summary, it appears that there is a trend towards sending less-hazardous waste off-site for treatment and disposal. It is possible that there is an associated reduction in waste generation, but no data are available to support this conclusion.

HEAVY METALS

In terms of environmental issues, heavy metals are a subset of persistent toxics—substances that retain their environmental impact for a relatively long period of time after release into the ecosystem. Since heavy metals are elements, they never truly decay; they can, however, change to different forms that may increase or decrease their toxicity. Additionally, they can disperse, accumulate, or undergo other physical changes leading to changes in the likelihood of exposure thus altering associated risk. Here, too, there is a dearth of future projections, but understanding current consumption and emissions of heavy metals can shed some light on trends.

As with all forms of pollution, heavy metals are released either as a primary product or a byproduct into the environment in a variety of ways, all of which generally fall into the categories of production, use, or disposal. Use and disposal losses can occur long after the toxics are produced, depending on the nature of the final product. For example, arsenic used in pesticides is considered to be released as the pesticides are applied, whereas releases associated with the decay and disposal of wood treated with arsenical preservatives occur over an extended period of time, up to thirty years (Loebenstein 1994, 9).

While the processes through which metals enter the environment are identical to those of other pollutants, their fate once there is very different. Because metals cannot decay, the amount in the ecosystem will continue to climb even after production and use have peaked. Environmental risks associated with heavy metals can remain long after their production and use have ceased. For example, since the element chromium does not decay, risk is reduced only through dispersion, sequestration, or conversion to a less toxic form. It is important to note that changes in form and transport do not always tend towards less risk over time. For example, the form of arsenic in aquatic systems can vary depending on the characteristics of the water, but the changes in form are reversible; thus, the toxicity of arsenic does not necessarily decrease over time (Spliethoff, Mason, and Hemond 1995, 2157–61).

As highlighted in Chapter 5, lead emissions to air have fallen 90 percent under the gasoline phase out and product bans. However, lead continues to be a concern in some areas due to its use in paint. Also, because lead (unlike chromium and some other elements) does not readily convert into its ionic form, it is quite stable. Thus lead that is released will remain in the upper layer of soil and "is only removed very slowly—over decades—from this upper layer" (von Moltke 1987, 29–30).

It is apparent that heavy metals will continue to be an environmental concern in the foreseeable future. It therefore may be useful to consider some trends in heavy metal production and use to understand how emissions are changing due to the pollution control system. Trends in chromium, mercury, and arsenic are discussed here. These elements were selected partly because of the availability of data and because they illustrate a variety of potential trends in heavy metal usage. We have not chosen a random sample, so any generalizations must be made with caution. The actual trend in usage of a particular metal depends on a number of factors. Obviously, the demand for goods that contain the element will affect usage, as will the availability of substitutes. The use of potential substitutes will be affected by costs and environmental or other policies that encourage substitution.

Chromium

Chromium is used in three basic industries: in the metallurgical sector as a component of stainless steel and other alloys and for plating; in the manufacture of chemicals such as plating compounds, corrosion inhibitors, and other chemicals; and as a component of high temperature refractory (a ceramic used in foundry and furnace construction). U.S. Bureau of Mines projections estimate usage of chromium over the 1993 to 2002 time period to be 13 percent lower than that over the 1983 to 1992 time period. In addition to decreased total demand, over the last twenty years there has been a change in industry structure, with share of total chromium demand for metallurgical uses increasing and the shares for chemical and refractory falling (Papp 1994, 88). Since total releases of chromium plus transfers, as defined by the Toxics Release Inventory (TRI), as a fraction of apparent consumption is lower for the metallurgical sector than for either the chemical or refractory sectors, the result should be a lower emissions rate of chromium per ton of apparent consumption (Papp 1994, 89). Regulations eliminating the use of chromium-based corrosion inhibitors in cooling towers and those regarding plating could explain part of the decline of chromium used in the chemical industry. Data on apparent consumption, industry structure, and chromium industry releases are given in Table 5. (There is no mining of chromium-bearing ore in the United States, and hence no emissions due to mining.)

Using historical TRI data and projected consumption, the Bureau of Mines estimated chromium emissions (releases plus transfers) over the 1993 to 2002 time period also to be 13 percent lower than those of the 1983 to 1992 period (Papp 1994, 89). This estimate of the decrease does not take into account the change in industry structure and improved pollution control technology and practices, both of which should lead to lower emissions rates (Papp 1994, 89). Because the risk associated with emissions will depend on a number of factors in addition to the quantity released (such as the toxicity of chromium species and exposure pathways), environmental quality will not necessarily be improved; however, the trend is encouraging, as lower emissions will lead to less accumulation of chromium in the environment. It should be noted that these emissions are relative indicators rather than absolute, because they are based

Table 5. U.S. Chromium Apparent Consumption and Releases Plus Transfers.

	1973– 1982	1983– 1992	1993–2002 (projected)
Average chromium apparent consumption (thousand metric tons annually)	459	400	349
Releases and transfers (metric tons annually)	28,241	24,611	21,447
Industry share of apparent consumption (percentage)			
Metallurgical	79	87	90
Chemical	12	10	9
Refractory	9	3	1

Source: Papp 1994.

on TRI data that only include releases from manufacturers and exempt certain activities and industries—including fossil fuel combustion and incineration (Papp 1994, 42). Fossil fuel combustion alone has been estimated to account for approximately 25 percent of the total chromium emissions in the United States (Papp 1994, 37–38).

Mercury

Mercury is another metal with industrial importance. Like chromium, mercury can exist in several forms, with varying toxicity. The Bureau of Mines breaks down mercury consumption into three general categories: chemical production (mostly through the chlor-alkali process), electrical and electronic uses, and instruments and related products. Trends of mercury use over the last fifteen years are similar to those of chromium: overall demand has fallen at the same time industry structure is changing. Between 1980 and 1990, total U.S. industrial consumption of mercury fell 65 percent; between 1990 and 1994, it fell an additional 33 percent. At the same time, electrical uses accounted for a decreasing share of mercury demand, falling from 54 percent of total demand in 1980 to 23 percent in 1994 (Jasinski 1994, 11; 1995). Most of this change is due to decreased use of mercury in batteries. These data are summarized in Table 6.

Some of the changes in mercury usage appear to be influenced by environmental regulation. Approximately 30 percent of chlor-alkali production capacity added in the United States between 1945 and 1970 was based on mercury cell technology, resulting in a fortyfold increase in the amount of chlorine that could be produced using this process. Since 1970, no mercury cell plants have been built due to environmental concerns with the mercury used. As of 1995, only about 12 percent of U.S. chlorine production came from mercury cell processes, down from 27 percent in

Table 6. U.S. Mercury Apparent Consumption.

	1980	1990	1994
Apparent mercury consumption (metric tons)	2,033	720	483
Industry share of consumption (percentage)			
Chemical	33	45	38
Electrical	54	29	23
Instruments	12	26	39

Source: Jasinski 1994.

1970 (Chlorine Institute 1995, Table 15). Not surprisingly, the consumption rate (that is, the amount lost from process per unit of production) of mercury fell from 0.12 kg Hg/mt Cl produced to 0.04 kg/mt between 1974 and 1994. The Chlorine Institute attributes this change to a decline in the number of plants using mercury cells and increasingly stringent environmental regulations (Jasinski 1994, 16).

In addition, mercury use in paint has dropped dramatically due to environmental concerns and policy actions. In particular, EPA, in cooperation with the domestic paint industry, banned the use of mercury in interior latex paint in 1990. At the same time many manufacturers voluntarily removed mercury from exterior latex paint. The result was a 90 percent reduction in consumption of mercury for paint between 1989 and 1990 (Jasinski 1994, 17). Environmental regulation has also led to the decrease of mercury contained in batteries. Due to concerns with the concentration of mercury in landfills arising from the disposal of batteries, several states enacted legislation that calls for the elimination of mercury in batteries sold within their borders (Jasinski 1994, 17). Presumably because of this type of pressure, the consumption of mercury in batteries fell nearly 60 percent in 1990 (Jasinski 1994, 17).

Overall mercury emissions projections are unavailable but are likely to fall with reduced demand. Emissions projections under a number of scenarios show 65 percent reductions of mercury released from MSW incinerators, due to the reduction of mercury in batteries and other products. Further reductions could be achieved with emissions controls and/or separation of mercury-containing waste from the waste stream (Shaub 1993, 52–54). As with chromium, decreasing mercury emissions do not necessarily imply reduced risk; but as they fall, lower rates of accumulation will occur in the environment. As with the chromium data, these numbers should be used as a relative comparison, as they do not include all sources of emissions. In particular, the single largest anthropogenic source of mercury emissions is coal combustion, which in 1990 accounted for approximately 12 percent of anthropogenic emissions to the environment (Jasinski 1994, 4, 24–25).

Table 7. U.S. Arsenic Apparent Consumption.

	1980	1990
Apparent arsenic consumption (metric tons)	12,400	20,500
Industry share of apparent consumption (percentage)		
Agricultural chemicals	46	20
Glass	5	4
Wood preservatives	44	70
Nonferrous alloys and electronics	3	4
Other	2	2

Source: Loebenstein 1994.

Arsenic

Arsenic is a metal with many different uses. Historically, it has been an important component of agricultural chemicals (primarily insecticides, herbicides, and desiccants to assist in the harvesting of cotton) and as a feed additive. Nonagriculturally, arsenic is used in wood preservatives, as an additive to certain metal alloys, and in semiconductors. In contrast to chromium and mercury, arsenic demand in 1989 was 80 percent greater than in 1980 (Loebenstein 1994, 9). Like the other two metals, the industry structure of arsenic changed over this time period. In 1980, agricultural products (primarily pesticides and desiccants) accounted for about 46 percent of total demand. By 1990, this had dropped to about 20 percent, while use as a wood preservative had increased to about 70 percent of total demand (Loebenstein 1994, 4–6, 9). The Bureau of Mines attributes the decrease in demand for arsenical pesticides to the introduction of organic pesticides (begun in the late 1940s) and EPA's actions, such as its preliminary decisions to cancel the registration of inorganic arsenical pesticides and arsenic-containing cotton desiccants (Loebenstein 1994, 4). Use of arsenical wood preservatives increased rapidly after 1975, primarily due to increased demand for these products in the residential sector (Easterling 1996; Loebenstein 1994, 6). Apparent consumption data are given in Table 7.

Emissions of arsenic will change as total demand changes. Additionally, the nature of arsenic emissions is more sensitive to recent changes in industry structure than are chromium and mercury. As a result of the phase out of arsenic as a component of agricultural chemicals (which are dissipative uses), it is likely that the near-term emissions rate for a given demand is probably lower now than in the past. (Because arsenic is being incorporated into products that have lifespans of twenty years or more, lower emission rates due to structural demand changes may only be a temporary effect. Ultimately the arsenic will probably enter the environment.)

Like the other two metals examined, using production and/or consumption figures as a proxy for emissions trends can be misleading, as significant sources of arsenic are not included in these figures. In 1989, copper smelting alone accounted for approximately 48 percent of arsenic emissions, mostly through slag and tailings disposal (Loebenstein 1994, 6–7).

While generalizations made from the examination of these three metals will be weak, some themes can be identified. First, because of the large portion of emissions of metals that are byproducts of other processes (such as fossil fuel combustion), production and consumption trends of metals do not tell the whole story. Furthermore, as emissions from production and consumption processes fall, the share of emissions as byproducts will tend to rise, all else being equal. Thus, using production and consumption trends as a proxy for emissions will become less valid without concomitant changes in emissions rates of a metal as a byproduct of other processes.

Secondly, the accumulation of metals in the environment will continue, even as emissions fall. As man mobilizes more metal from within the earth, that metal will eventually reach the ecosystem, unless it is continually recycled or permanently sequestered. Until the loop is completely closed and mining of metals ceases, they will continue to accumulate in the environment. While this build-up does not always translate into higher (or significant) risk, in many cases it will.

The pollution control system appears to have had an effect on the way certain metals have been used in our economy, although it is not clear how much of the observed structural changes can be directly attributed to these policies. This apparent effect may illuminate ways to move toward a system based on pollution prevention, the practice of eliminating the creation of waste and/or unwanted toxic materials, as opposed to simply treating effluents to prevent release of these contaminants into the environment. In the case of mercury and chromium, it appears that these changes will result in fewer absolute emissions and, in the case of arsenic, it is possible that the rate of emissions is lower than it would have been without the pollution control system. A still unresolved question is whether these changes will result in lower risk.

NEW ITEMS ON THE AGENDA

Thus far, we have explored the future by extending the trends of current environmental issues. The other class of future issues includes those items that will be new on the environmental agenda. While many of these items will be unforeseen, the outlines of some are already apparent.

In 1995, EPA's Science Advisory Board issued a report that, in part, identified some long-term environmental issues that will be new on the agenda (U.S. EPA 1995a). The five broad problem areas identified were the sustainability of terrestrial ecosystems, noncancer human health effects, total air pollutant loadings, nontraditional environmental stressors, and the health of the oceans. Each of these areas is beginning to gain prominence both within the environmental community and for the general public. Focusing on the potential of manmade chemicals to mimic animal and human hormones and subsequently damage reproductive systems, the book *Our Stolen Future*, for example, has generated controversy over its methodology and conclusions. Whether the book's findings are ultimately accepted or rejected, its prominence shows that the impact of manmade substances on reproductive systems (a noncancer human health effect) is clearly on the environmental agenda.

Since concern for these five areas is relatively new, the pollution control system is just beginning to grapple with them. Whether it will effectively meet these future challenges depends on both the risk these items actually pose and how the system will address these risks. Thus, it is still too early to assess whether the pollution control system has performed adequately but the fact that new concerns were identified shows the system is able to anticipate some problems and thus reduce the number of future surprises.

Another objective of the Science Advisory Board's report was to identify ways the EPA could incorporate futures research into its activities—to institutionalize forecasting within the agency so the pollution control system can respond quickly and effectively to any new challenges. While forecasting alone does not guarantee that new issues will be successfully addressed, the pollution control system will likely handle new issues better with a futures research group than without one.

SUMMARY

While we may draw some tentative generalizations about future environmental conditions and challenges, they are closer in spirit to "themes" or "cautionary measures" than definitive conclusions, because the data on which they are based are incomplete. Perhaps the most important theme of this chapter is that environmental protection is not solely the domain of EPA. Many other national policies have major impacts on environmental quality. Energy and agriculture are just two examples of policies that have an enormous influence on current and future emissions trends, and they are both beyond the scope of the EPA. Until environmental concerns are truly integrated with the policies of other agencies, the possibility of other agencies' objectives working against pollution control remains.

Attempts have been made to better coordinate policy and programs among different agencies and levels of government, as exemplified by efforts to improve water quality through a watershed approach. These projects often involve several federal agencies as well as state and local levels of governments. While a step in the right direction, these programs face formidable institutional barriers. Chief among them are the different and sometimes conflicting missions of the agencies, and the inflexibility of federal activities, which often arises from the laws and regulations that guide the agencies (U.S. GAO 1995, 7). Flexibility, identified earlier as a precondition for addressing unforeseen environmental issues, is also necessary for effective policy coordination. In addition, without a solid commitment through funding and organizational support, coordination efforts will not succeed.

A second, related theme to emerge from this analysis is the tension between increased economic activity and increased environmental efficiency (the amount of pollution per unit of activity, whether measured by BTU, GDP or another metric). As has been shown, emissions of CO, VOC, and NO_x due to energy use fall between 1990 and 2000, primarily due to less polluting energy technologies. Beyond that time, however, increased consumption is projected to overcome the gains made by more environmentally efficient technologies. A similar pattern occurs in municipal solid waste disposal, where decreasing per capita disposal is overcome by increased population, leading to an absolute increase in quantities disposed. These observations temper the technology optimist perspective that low-emissions technologies will automatically lead to fewer environmental impacts. Benefits that are achieved through better technologies and lower emissions rates can be more than offset by increased intensity of use. As the economy and the population of the nation continue to grow (energy consumption forecasts assume an annual GDP growth of 2.2 percent between 1994 and 2010), pressures on the environment may increase, even as we pollute less on a per capita or per GDP basis.

This tension highlights two vastly different perspectives on the consequences of growth, and thus on future environmental conditions. At one extreme, some authors argue that exponential economic and population growth coupled with a constant carrying capacity of Earth will lead to severe environmental problems unless immediate action is taken to curb expansion. In 1971, for example, the Club of Rome sponsored a computer simulation of future world environmental conditions by scientists at the Massachusetts Institute of Technology (MIT) who summarized their findings in the book *The Limits to Growth* (Meadows and others 1972). The scientists factored in population, resources, production and pollution, assuming that rates of change for each of these variables was a linear function of the values of all the others. Akin to the doomsday assumptions of Malthus (1798), who wrote about collapse from overpopulation and famine nearly two centuries earlier, the now-much-maligned MIT model predicted global collapse within seventy years, if human patterns of population and economic growth failed to change. Like the Malthusian model, which failed to anticipate the agricultural revolution, the "limits" study failed to consider the degree to which technical change would avert such grim predictions.

On the other side of the "limits" argument are those who argue that the carrying capacity of Earth can be expanded through technological advances and structural changes in the economy (see for example Ausubel 1996 and Hodges 1995). While conceding that nonrenewable resources are, by definition, finite and that there probably exists an absolute limit to carrying capacity, these authors contend that human adaptability and ingenuity will allow growth to continue without severe repercussions, at least in the near term.

While the dire predictions of the "limits" group have largely been discredited, another wave of challengers to conventional economics, known as "ecological economists" have ushered in a new call to temper growth with environmental concerns. The new movement, identified with organizations such as the International Society for Ecological Economics and the Sweden-based Beijer Institute, is heavily influenced by the steady-state economics approach advanced during the late 1960s by such nontraditional economists as Kenneth Boulding and Herman Daly (see, for example, Daly 1973 and Pfaff 1976).

Ecological economics (see, for example, Costanza 1991), as well as a larger intellectual movement supporting "sustainable" development, was spurred in large part by the influential Brundtland report, *Our Common Future*. The report calls for a middle-of-the-road approach that aims to integrate environmental and ecological concerns (WCED 1987). Sustainable development rejects both extremes represented by the "limits" group and the technological optimists. Sustainability theorists focus instead on minimizing "throughput"—that is, the total flow of resources from nature to consumption and disposal. Traditional economic measures view increasing throughput as a "good" that signals prosperity. In contrast, the new sustainability camp, colored heavily by emerging insights from ecology, seeks to keep throughput beneath "carrying capacity"—the threshold above which Earth becomes unable to absorb or "carry" effects from pollution or resource extraction (Goodland 1991). While the sustainability approach moves away from the two extremes that predict either boom or gloom, its focus on "carrying capacity" may have the unintended result of merely shifting the debate away from uncertainties surrounding our ability to solve scarcity through technological innovation, towards even more unpredictable measures regarding the planet's capacity to absorb the effects of human activity.

In contrast to a global framework that characterizes the various intellectual future scenarios, our analysis has been limited to the United States, and attempts to say nothing definitive of the rest of the world. Based on the projections investigated, however, it seems unlikely that in the next twenty years the United States will face a severe environmental crisis caused by population or economic growth. This does not rule out the possibility of surprises, nor do we imply there will be no environmental issues of concern; but for the next fifteen to twenty years, the economic and population growth of the United States will probably not lead us over the environmental cliff. Even in this relatively short time frame, however, some environmental problems are likely to become worse because of economic and population growth. In a longer time frame it is impossible to know whether growth will outstrip our technological abilities to control its adverse effects.

Another future issue we have highlighted is the growing importance of nonpoint source pollution. Historically, the pollution control system has focused on point source pollution in part because these sources were easy to identify and regulate. Nonpoint source pollution is more difficult to regulate and control, as it usually results from numerous, geographically dispersed sources each emitting relatively small quantities of pollutants. Such pollution does not become an issue until these quantities accumulate in some part of the ecosystem. For the pollution control system to adequately manage these sources, it will somehow have to influence many actors to reduce their seemingly insignificant amounts of pollution. The historical approach of command and control is ill-suited to meet this challenge: expecting the federal government to control how much each person drives his or her car, for example, is infeasible from both resource and political standpoints. Given the current public distrust of government, any potentially intrusive action by a governing or regulatory body will likely be met with severe resistance by both the public and elected officials.

A final conclusion is the need for more data. It is logical for organizations to spend their resources mitigating the problems of today; without forecasting, however, opportunities to change the trajectory of the pollution control system in a positive direction will be missed (Kemp and Soete 1992, 439–41). The prevention of pollution today is likely to be less costly than the retrofit solutions of tomorrow.

The dearth of forecasting information points to an obvious need in the system. By working to identify the impacts and trends in environmental policy, we may be able to reduce if not eliminate the adverse consequences of future human activity on the natural world. Certain EPA programs (such as reregistration of pesticides and the Toxic Substances Control Act) require an understanding of the potential impacts of new chemicals on the environment before they enter our socioeconomic system, with the aim of preventing future problems. In addition, the Environmental Futures Committee of the Science Advisory Board and a small forecasting group have recently been created to identify future issues for EPA. While these are steps in the right direction, such efforts are the exception rather than the rule in the current pollution control system.

Reasonable forecasting requires an understanding of the current situation. As shown in Chapter 5, current monitoring data do not meet this requirement. The need to develop systems that can provide systematic, statistically valid evidence of trends will become more critical as pocketbooks are held tighter. In particular, given the rise in relative importance of nonpoint source pollution, projecting the behavior of many diverse actors is more pertinent now than ever, despite the methodological difficulties involved.

ENDNOTES

[1]While desalination of salt or brackish water is an additional source of fresh water, it accounted for less than 0.2 percent of U.S. withdrawals in 1990 (Gleick 1993, 422), and is not likely to become a significant source of supply in the next decade. Costs of desalination, while competitive in some locations, still generally exceed other potential sources.

REFERENCES

Ausubel, Jesse H. 1996. Can Technology Spare the Earth? *American Scientist* (March–April): 166–78.

Bernardini, Oliviero, and Riccardo Galli. 1993. Dematerialization: Long-Term Trends in the Intensity of Use of Materials and Energy. *Futures* 25(4, May): 431–48.

Chen, H., and A. D. Druliner. 1988. Agricultural Chemical Contamination of Groundwater in Six Areas of the High Plains Aquifer, Nebraska. In *National Water Summary 1986: Hydrologic Events and Groundwater Quality*. Water Supply Paper 2325. Washington D.C.: U.S. Geological Survey, 103–8.

Chlorine Institute. 1995. *North American Chlor-Alkali Industry Plants and Production Data Book*. Pamphlet 10. Washington, D.C.: Chlorine Institute.

Costanza, Robert., ed. 1991. *Ecological Economics: The Science and Management of Sustainability*. New York: Columbia University Press.

Costanza, Robert, Lisa Wainger, Carl Folke, and Karl-Goran Maler. 1993. Modeling Complex Ecological Economic Systems. *BioScience* 43(8): 545–55.

Daly, Herman, ed. 1973. *Toward a Steady-State Economy.* San Francisco: W. H. Freeman.

Easterling, Jeff. 1996. Personal communication with authors, March 8. (Jeff Easterling is with the Southern Forest Products Association).

Frederick, Kenneth. 1992. Managing Water for Economic, Environmental, and Human Health. *Resources* (106, Winter): 22–25.

———. 1995. An Overview: America's Water Supply: Status and Prospects for the Future. *Consequences* 1(1, Spring): 13–23.

Gleick, Peter H., ed. 1993. *Water in Crisis: A Guide to the World's Fresh Water Resources.* New York: Oxford University Press.

Goodland, Robert, ed. 1991. *Environmentally Sustainable Economic Development: Building on Brundtland.* Paris: UNESCO.

Guldin, R. W. 1989. *An Analysis of the Water Situation in the United States: 1989–2040.* U.S. Forest Service. Fort Collins, Colorado: USDA.

Hodges, Carroll Ann. 1995. Mineral Resources, Environmental Issues, and Land Use. *Science* 268(June 2): 1305–12.

Jasinski, Stephen M. 1994. *The Materials Flow of Mercury in the United States.* U.S. Bureau of Mines Information Circular 9412. Washington, D.C.: U.S. GPO.

———. 1995. Personal communication with the authors.

Karagozoglu, Necmi. 1993. Environmental Uncertainty, Strategic Planning and Technological Competitive Advantage. *Technovation* 13(6): 335–47.

Kemp, Rene, and Luc Soete. 1992. The Greening of Technological Progress. *Futures* (June): 24: 437–57.

Loebenstein, J. Roger. 1994. *The Materials Flow of Arsenic in the United States.* U.S. Bureau of Mines Information Circular 9382. Washington, D.C.: U.S. GPO.

Malthus, Thomas. 1798. Essay on the Principle of Population. As referenced in Robert L. Heilbroner. 1953. *The Worldly Philosophers: The Lives, Times, and Ideas of Great Economic Thinkers.* New York: Simon and Schuster.

Meadows, D. H., D. L. Meadows, J. Randers, and W. W. Beherns. 1972. *The Limits to Growth: A Report for the Club of Rome's Project on the Predicament of Mankind.* New York: Potomac Associates and Universe Books.

Papp, John F. 1994. *Chromium Life Cycle Study.* U.S. Bureau of Mines Information Circular 9411. Washington, D.C.: U.S. GPO.

Pfaff, Martin., ed. 1976. *Frontiers in Social Thought: Essays in Honor of Kenneth E. Boulding.* Amsterdam: North-Holland.

Placet, M., D. G. Streets, and others. 1987. *National Acid Precipitation Assessment Program (NAPAP) Interim Assessment: The Causes and Effects of Acidic Deposition: Volume 2. Emissions and Controls NAPAP.* Washington, D.C.: NAPAP.

Repa, Edward W., and Allen Blakely. 1996. *Municipal Solid Waste Disposal Trends: Update.* Washington, D.C.: National Solid Waste Management Association.

Shaub, Walter M. 1993. Mercury Emissions from MSW Incinerators: An Assessment of the Current Situation in the United States and Forecast of Future Emissions. *Resources, Conservation and Recycling* 9: 31–59.

Smith, Richard A., Richard B. Alexander, and M. Gordon Wolman. 1987. Water-Quality Trends in the Nation's Rivers. *Science* 235: 1607–15.

Solley, W. B., R. R. Pierce, and H. A. Perlman. 1993. *Estimated Use of Water in the United States in 1990.* U.S. Geological Survey Circular 1081. Washington, D.C.: U.S. Geological Survey.

Spliethoff, Henry M., Robert P. Mason, and Harold F. Hemond. 1995. Interannual Variability in the Speciation and Mobility of Arsenic in a Dimictic Lake. *Environmental Science and Technology* 29(8, August): 2157–61.

USDA (U.S. Department of Agriculture). 1994. *Agricultural Resources and Economic Indicators.* Agriculture Handbook 705. Economic Research Service. Washington, D.C.: USDA.

U.S. DOE (Department of Energy). 1993. *Emissions of Greenhouse Gases in the United States 1985–1990.* Energy Information Administration. Washington, D.C.: U.S. DOE.

———. 1994a. *Annual Energy Review 1993.* Energy Information Administration. Washington, D.C.: U.S. DOE.

———. 1994b. *Emissions of Greenhouse Gases in the United States 1987–1992.* Energy Information Administration. Washington, D.C.: U.S. DOE.

———. 1995. *Annual Energy Outlook 1995.* Energy Information Administration. Washington, D.C.: U.S. DOE.

U.S. EPA (Environmental Protection Agency). 1994a. *National Capacity Assessment Report: Capacity Planning Pursuant to CERCLA Section 104(c)(9).* Draft. Office of Solid Waste and Emergency Response. Washington, D.C.: U.S. EPA.

———. 1994b. *National Water Quality Inventory: 1992 Report to Congress.* Office of Water. Washington, D.C.: U.S. EPA.

———. 1995a. *Beyond the Horizon: Using Foresight to Protect the Environmental Future.* Science Advisory Board. Washington, D.C.: U.S. EPA.

———. 1995b. *EPA National Air Pollutant Emission Trends, 1900–1994.* Office of Air Quality Planning and Standards. Research Triangle Park, North Carolina: U.S. EPA.

———. 1996. *Characterization of Municipal Solid Waste in the United States: 1995 Update.* Office of Solid Waste and Emergency Response. Washington, D.C.: U.S. EPA.

U.S. GAO (General Accounting Office). 1995. *Agriculture and the Environment: Information on and Characteristics of Selected Watershed Projects.* Washington, D.C.: U.S. GAO.

USGS (U.S. Geological Survey). 1988. *National Water Summary 1986: Hydrologic Events and Groundwater Quality.* USGS Water Supply Paper 2325. Washington D.C.: USGS.

von Moltke, Konrad. 1987. *Possibilities for the Development of a Community Strategy for the Control of Lead.* Unpublished report on file with authors.

WCED (World Commission on Environment and Development). 1987. *Our Common Future.* New York: Oxford University Press.

PART III:
Overview

11

Conclusions

What conclusions can we draw from this journey through the complexities of the pollution control regulatory system? Four general themes emerge.

First, the fragmented system is seriously broken. Its effectiveness in dealing with current problems is questionable, it is inefficient, and it is excessively intrusive. These are fundamental problems.

Second, the problems cannot be fixed by administrative remedies, pilot programs, or other efforts to tinker at the margins. They are problems that are built into the system of laws and institutions that Congress has erected over thirty years. We recognize the difficulty of ever achieving fundamental, nonincremental change in the American government, but nothing short of such change will remedy the problems we have identified.

Third, the picture is not all bleak. The system has accomplished some solid victories in the quest for environmental quality. Some elements of the existing system have worked well. To cite just one, the efforts of EPA to incorporate citizens' views and to encourage openness and accessibility have been pioneering and constructive. In trying to improve the system, care must be taken to preserve the positive aspects of it.

Fourth, a dearth of information of all kinds characterizes pollution control. The system lacks monitoring data to tell whether environmental conditions are getting better or worse; it lacks scientific knowledge about both the causes and the effects of threats to human and environmental health; it lacks information that would tell us which programs are working and which are not. There is a systematic underinvestment, in some cases a disinvestment, in the information-gathering efforts necessary to run an intelligent pollution control program.

In the past decade views of government officials, elected representatives, the media, and the public have increasingly diverged about whether the U.S. pollution control regulatory system is performing satisfactorily. Some people point to the significant reductions in most air pollutants, the cleanup of major rivers, and the tangible progress that has been made in improving environmental quality. Others point to the inefficiency and intrusiveness of regulations, and the lack of progress in dealing with nonpoint pollution sources or global climate change. The general public continues to strongly support pollution control, but at the same time it often opposes the taxes necessary to pay for it and mistrusts the government officials who administer it.

A review of our findings, based on the criteria we used to evaluate the system, helps to explain the divergent views about the system. The six criteria we used are:

- Has the system reduced pollution levels?
- Has it targeted the most important problems?
- Has it been efficient?
- How responsive has it been to a variety of social values (such as public involvement, nonintrusiveness, and environmental justice)?
- How does it compare with systems in other developed nations?
- How well can it deal with future problems?

REDUCING POLLUTION LEVELS

We began by looking at actual levels of pollution in the environment and how these levels have changed over time. This is not an easy task, because, for almost every type of pollution, monitoring data are woefully inadequate. We can tell something about national air pollution trends, although the data are often sparse. This is better than we can do for water quality, where it is impossible to draw any firm conclusions on a national or state basis about whether water quality is improving or getting worse. Most other areas are like water quality in that they lack the data necessary to know whether conditions are improving over time.

Overall, it seems fairly certain that the environment generally is better in 1997 than it was in 1970. It is certainly better than it was in 1940 or 1900, although such comparisons cannot be based on hard data and also involve subjective comparisons (such as between the effects of horse manure and diesel fumes). The regulatory system has prevented both population growth and huge increases in the U.S. economy from exerting large negative environmental effects.

Data for particular environmental areas are summarized below.

Air Quality

EPA concentrates its attention on six major air pollutants, called criteria air pollutants. Four of the six have shown significant improvement, both historically and in recent years. Between 1986 and 1995, average airborne concentrations of lead dropped 78 percent, while those of carbon monoxide and sulfur dioxide dropped 37 percent (U.S. EPA 1996f, 10, 14, 19, 26). Concentrations of particulate matter also have improved, although not as much. The regulatory definition of particulate matter changed in 1987 to focus on smaller particles, so long-term trend data are not available. The two criteria pollutants that have not shown as much improvement—nitrogen dioxide and ozone—are related to urban smog (U.S. EPA 1995a, ES-8; 1996f, 1).

The general improvement in air quality still has left some areas of the nation experiencing days when air pollution exceeds national health-based standards. Ozone, an indicator of smog, accounts for most of the failures to attain the standards. In 1996, approximately 127 million people lived in counties classified as nonattainment areas for one or more criteria pollutants (U.S. EPA 1994a, Figure 5-9). It is important to note, however, that a county is considered to not meet a standard even if the standard is only violated once or twice a year at only one of possibly several monitors in the area; the figure in this sense may exaggerate the degree of the problem.

Water Quality

It is not possible to piece together a coherent picture of trends in U.S. water quality based on existing data. Overall, it appears likely that the quality of the nation's rivers and lakes has improved dramatically in a few places but improved more modestly or stayed the same in most others. Water quality in estuaries and coastal waters probably has declined.

EPA's 1994 water quality report (EPA issues such reports biennially, based on state data) concludes that 43 percent of assessed rivers and 50 percent of assessed lakes fail to meet Clean Water Act goals that they be fishable and swimmable (U.S. EPA 1995c, ES-13, ES-16). This statement may exaggerate the extent of the water pollution problem, because fewer than 20 percent of the rivers and only half the U.S. lakes were assessed; probably the more polluted ones received assessment.

The National Stream Quality Accounting Network (NASQAN), run by the U.S. Geological Survey, provides the only consistent data on water quality trends. The network data, while quite reliable, do not deal with most of the pollution situations that EPA currently worries about, such as oxygen demand from effluents around urban areas. The data show mixed trends in the 1974–81 period and generally improved water quality in the 1980–89 period (Smith, Alexander, and Wollman 1987, 4; Smith, Alexander, and Lanfear 1991, 117–27).

Looking at the sources of water pollution, EPA can claim one notable victory and must acknowledge one notable gap. The victory is in treatment of household or municipal sewage. For most of its history, EPA's largest program in dollar terms provided grants to local governments (through state governments) to build municipal waste treatment plants. In a 1988 evaluation to Congress, EPA reported that the number of people in communities receiving secondary or advanced levels of treatment rose from 4 million in 1960 to 143.7 million in 1988 (U.S. CEQ 1990, 309). Since 1960, the number not served at all by public treatment facilities has hovered around 70 million. Some of the facilities that exist are poorly operated, but the massive increase in areas served by secondary treatment is a major accomplishment.

The glaring gap in EPA's water quality program is nonpoint sources, now the leading source of water pollution in most areas. Nonpoint sources include runoff from farms and city streets and also deposition of pollutants from the air into water bodies. While this type of pollution is more difficult to control than pollution from discrete outfall pipes at, say, factories or municipal treatment plants, a variety of measures could be taken. Neither Congress nor EPA has done much to address nonpoint sources, in part because of reluctance to take on the farm lobby. Since 1990, the $95 million per year EPA has given to states to deal with nonpoint sources is less than 5 percent of the total funds it gives for water quality (U.S. EPA 1996h).

Drinking Water

The law governing federal efforts to assure the safety of drinking water was amended in 1996. Data on the effectiveness of the old program do not tell us much. The percentage of drinking water systems in compliance with federal standards hovered between 68 and 73 percent from 1986 to 1993 (U.S. EPA 1995b, 22). The number of violations rose and fell sharply, but the changes were due more to changed reporting requirements than to any changes in actual water conditions. Prevention of drinking water contamination is likely always to be more a function of local resources and efforts than of the federal program, although the new federal law contains a fund to assist localities in improving their drinking water facilities.

Municipal Solid Waste

The amount of garbage each of us generates has doubled since the 1960s (U.S. EPA 1996d, 67). In large part, the amount reflects economic prosperity: people are buying, consuming, and discarding more things. The total amount of garbage generated has increased even more because there are more people producing it.

EPA encourages recycling by households, but local governments and private markets have the most influence on how much material is actually recycled. EPA regulates the safeguards that landfills and incinerators must employ, but the risks arising from these sources are not easy to detect, nor do good data exist on compliance of the facilities with the regulations. A significant accomplishment that can be credited to the Resource Conservation and Recovery Act (RCRA) is the virtual elimination of open burning of garbage, a widespread practice before the 1970s.

Hazardous Waste

Despite RCRA's broad tracking and reporting provisions for hazardous waste, few data exist to show whether the law is achieving its goals (EESI 1987; Portney 1990, 168). What we do know is that whereas only a few years ago there was talk of a "crisis" in capacity for hazardous waste disposal sites, a surplus of disposal capacity now exists. What we do not know are the reasons for the surplus. It is likely due to more of the waste being treated on-site and to an increase in incinerators rather than to an actual reduction in the amount of waste generated. Whether this is a good trend for the environment is unclear. The RCRA program almost certainly has improved the methods used for handling hazardous waste. For example, land disposal of untreated hazardous waste has been greatly reduced.

The second largest EPA program (after water) as measured by direct program dollars spent is the so-called Superfund program, designed to clean up sites contaminated by hazardous substances (NAPA 1995, 20). The program has been a subject of controversy almost from its beginning. The major risks from hazardous waste sites have probably been addressed through emergency removal actions; 3,042 such actions have been undertaken as of 1995 (U.S. EPA 1996i). Permanent cleanup of Superfund sites has proceeded more slowly, in part because negotiating the remedies to be applied and who should pay for them can be a slow rancorous process and in part because the actual cleanup, especially of contaminated groundwater, can take many years. As of June 1996, construction work had been completed at 410 of the 1,227 sites on the Superfund National Priorities List (U.S. EPA 1996i).

Toxics

More than 2,000 new chemicals are reviewed each year by EPA through an EPA review process that is generally credited with significantly reducing the toxicity of new chemicals manufactured (U.S. EPA 1996b, 31); unfortunately, no data exist to support the view. Although the manufacturer of a proposed new chemical is required to submit the results of any tests conducted on the chemical, less than 10 percent of the new chemical notifications sent to EPA contain any test data (U.S. EPA 1997).

Seventy thousand chemicals are currently manufactured or processed in the United States (Reid 1992, 30). EPA has banned five substances under the Toxic Substances Control Act. The agency now collects emissions data on about 600 chemicals (282 new chemicals and chemical categories were added in 1995) under the Toxics Release Inventory, but because the TRI data cover only some emissions sources, the inventory underestimates emissions (U.S. EPA 1997b; Box 4-E). Total TRI figures are complicated at best because of changes in chemicals and industries covered. The TRI data are divided between direct releases to the environment and transfers off-site. Direct releases have been reduced by about half between 1988 and 1995 but transfers have more than doubled (U.S. EPA 1996b, Table 5-2).

In 1991, EPA targeted seventeen of the most toxic substances on the TRI list for reduction under a voluntary program called "33/50." The program achieved its goal of a 50 percent reduction in *emissions* for these chemicals (U.S. EPA 1996a, 173). Long-term *production* figures for these seventeen chemicals show a mixed picture, however. Between 1970 and 1990, production declined for three of the chemicals, increased for five, and was not significantly changed for six (U.S. ITC 1970–1990, 11-1, 15-1; U.S. DOI 1970–1990). Data were not available for three of the chemicals. The government has stopped collecting data on chemical production because of budget reductions.

The fragmentary data on the level of toxics found in humans and wildlife indicate that significant progress has been made in addressing some persistent toxics. The average level of lead found in the blood of humans has declined 78 percent from 1976–1980 to 1988–1991 (Dodell 1994). The level of PCBs, long-lived chlorinated compounds that may cause cancer, has declined in Lake Michigan trout from a peak of twenty-three in 1974 to less than three in 1990 (measured in micrograms per gram wet weight) (U.S. CEQ 1993, 491). Similar reductions have been achieved for other chlorinated compounds (such as DDT) and in other locations.

Pesticides

Pesticides, including fungicides, herbicides, and similar products, are a unique group of chemicals because they are intended to be toxic to some living things and they are deliberately introduced into the environment. Because of this combination they are closely regulated, and EPA reviews and approves every pesticide for every particular use.

More than half of pesticide use is on agricultural crops, so the volume of pesticides used tends to vary with the volume of agricultural production, the crop mix, and the technologies used in agriculture (U.S. EPA 1991b, 2). Total pesticide use in agriculture increased from 232 million pounds in 1964 to 573 million pounds in 1992 (USDA 1994, 86). However, the peak year was 1982 (612 million pounds). The mix of pesticides used has changed significantly. Between 1964 and 1992, use of insecticides declined by 60 percent while use of herbicides increased by more than 600 percent (USDA 1994, 86). Total U.S. production of pesticides has remained at about 1.1 billion pounds for each of the past ten years (USDA 1994, 86).

As noted above, the level of persistent toxics, including pesticides, in humans and wildlife has declined significantly. Newer pesticide compounds are less persistent in the environment. They tend to be more acutely toxic than the older types of pesticides but are less likely to cause chronic problems such as birth defects. The revolution in genetic engineering is likely to have a major impact on the amount and types of pesticides used in the future, but it is too early to know the environmental implications of this change.

Problems Determing Cause and Effect

How much of any change in pollution levels is due to pollution control regulatory programs? We know that both weather (especially temperature and precipitation) and economic changes directly influence pollution levels apart from any pollution control efforts. In an effort to statistically disentangle these factors, we analyzed economic, meteorological, and air pollution control cost data for a twenty-year period in three metropolitan areas where the level and composition of economic activity substantially changed concurrently with the implementation of national air quality standards: Allegheny, Pennsylvania (which includes the city of Pittsburgh); Baltimore, Maryland; and Cuyahoga, Ohio (Cleveland). Air quality in these areas has improved dramatically over the past few decades, but the improvement is often attributed to the decline in heavy manufacturing, leaving the impression that mandated pollution control investments have had little impact. Previous studies have found little evidence of an association between pollution control investments and air quality.

The results of our analysis suggest that mandated pollution control investments often have had a significant effect on reducing air pollution levels. Because we found such an effect in areas where economic changes have been stark, it is reasonable to assume that significant regulatory effects also have occurred where changes in the level and composition of economic activity have been less dramatic. However, in the three areas we examined, the effects of economic changes, weather, and other factors were much greater than the effects of regulatory controls. Also, the failure of local factors (such as the level of local manufacturing activity and local pollution control investments) to account for a majority of the variation in local air quality underscores the importance of regional or national factors (both regulatory and nonregulatory) in determining local air quality.

TARGETING THE MOST IMPORTANT PROBLEMS

Regardless of how effectively EPA is reducing the levels of pollutants on which it is focused, if it is focusing on the wrong targets then it may not be doing much to address protection of human health and the natural environment—the real goals of the pollution control regulatory system. Investigating this question takes us into the realm of priority-setting and comparative risk.

Although we talk about EPA's priorities, the priorities are set primarily by Congress. As we noted earlier, statutes drive pollution control priorities, and EPA has limited flexibility to adjust legislatively established priorities. Furthermore, setting priorities is dependent on other factors in addition to the risk presented by the problem. The cost of control measures, their implementability, and how the costs and benefits are distributed are just some of the other dimensions that determine program priorities. However, the relationship between priorities and risk is a fundamental criterion that should be considered in evaluating the pollution control system.

The best indicator of EPA's priorities is the money spent on particular problems. Most analyses have focused on EPA's budget expenditures, but this tells only part of the story: the real impact of EPA programs often depends on expenditures by state and local governments and the private sector. The private sector bears about 60 percent of total pollution control costs; local governments account for 20 to 25 percent of total costs; EPA and other federal agencies account for less than 10 percent (U.S. EPA 1991a, 8-51). States supply the remainder.

EPA's general budget priorities have remained fairly constant since the agency was created. Water pollution control has topped the list, followed by air pollution and hazardous waste programs. At this general level, private sector spending priorities are the same as EPA's. However, at the level of specific programs, disparities between public and private spending priorities appear. For example, the Superfund program in fiscal year 1994 accounted for 26 percent of EPA's budget but only 3 percent of projected total U.S. pollution control expenditures (U.S. OMB 1996, 869–82; U.S. EPA 1990, 3-3, 4-3, 5-3, 6-3, 7-3, 8-3). Few reliable data exist on the actual cost of Superfund site cleanup.

The first comparative assessment of the risks controlled by EPA programs was performed in 1987 by a team of senior EPA analysts (U.S. EPA 1987). Since then, EPA's Science Advisory Board has conducted two extensive reviews of the 1987 report, each of the ten EPA regions has performed a comparative risk assessment, and twenty-five states and several localities have compiled risk rankings (U.S. EPA 1993). The methods used by the states varied widely, and the rankings were done at different times. On the whole, however, the results were quite similar.

Comparing these rankings with expenditures by either EPA or society at large reveals a huge gap between risk priorities and spending priorities. Indoor radon, which tops the health concerns of EPA regional officials, is not regulated by EPA at all and only minimally by the states. The EPA radon program is 0.07 percent of the EPA budget (U.S. EPA 1996g). Similarly, indoor air pollution, the second-ranked health problem, is not for the most part regulated by EPA or the states (U.S. EPA 1996g). Local building codes address some aspects of the indoor air problem, such as rates of air exchange. EPA has an indoor air program—it is 0.24 percent of the agency's budget (U.S. EPA 1996g).

The story on ecological risks is not much better. Nonpoint source pollution is rated the most serious ecological risk; to the extent that it is regulated at all, it is regulated by the states. The EPA nonpoint program accounts for 1.7 percent of its total budget (U.S. EPA 1996h). Stratospheric ozone depletion, the number two ecological risk, is being regulated; in fact, EPA's regulation of aerosol propellants was the pioneering move in addressing the stratospheric ozone problem. Pesticides, which rank third on both the health and ecological risk ranking, are regulated by both EPA and the states. The EPA pesticides program, however, receives only 2.4 percent of the EPA budget (U.S. OMB 1996, 869–82; U.S. EPA 1990, 3-3, 4-3, 5-3, 6-3, 7-3, 8-3).

All questions of comparative risk are plagued by the inadequacy of information about the nature and severity of environmental problems. There are not enough toxicity data on most chemicals to know whether they cause adverse effects. There are not enough monitoring data to know to which pollutants people are exposed. We do not understand many fundamental aspects of the earth's ecology—we do not understand the role of clouds in the Earth's temperature balance or what makes flowers bloom. Knowledge about how pollutants travel from one part of the environment to another is woefully inadequate. These are problems both of fundamental scientific knowledge and of inadequate data collection.

Another difficulty with comparative risk is the effect of past and existing control efforts. Drinking water, for example, would likely be the major environmental health threat were it not for existing water purification efforts. Nonpoint sources of stream pollution are so important now in part because of past success in controlling point sources. The results of comparative risk analyses therefore should be used primarily as guidance for allocating future efforts, not as a justification for dismantling current efforts.

As we have noted, comparative risk is not the only relevant factor in determining expenditure levels or priorities. The potential effectiveness of government action, the cost of dealing with a problem, and numerous other factors are important to consider. However, the conclusion seems inescapable that the pollution control regulatory system is not addressing some very important problems. It is focusing almost exclusively on outdoor air pollution when a large part of the health risk comes from indoor air pollution. It is focusing on point sources of water pollution when the major problem today is nonpoint sources. The current priorities are driven by the pollution control statutes. Until these statutes are changed, the priorities will be askew.

EFFICIENCY

The economic analysis of environmental regulation is complicated by at least two major problems. First, it is very difficult to evaluate either the costs or the benefits of environmental programs and regulations. Second, most of the benefits of environmental regulations are excluded from the standard macroeconomic indicators: cleaner air, for example, is not a gain when calculating gross domestic product.

The several attempts over the past fifteen years to estimate the aggregate benefits and costs of the Clean Air and Clean Water acts suffer from significant limitations, but their findings are consistent. Taken as a whole, the benefits of the Clean Air Act seem clearly to outweigh the costs. Three separate studies using different methods and different data sources conclude that aggregate benefits exceed aggregate costs. One study that looked just at the EPA rules promulgated between 1990 and 1995 concluded that the benefits of the Clean Air Act rules in this period exceeded the costs by about $70 billion in 1994 dollars (U.S. EPA 1996c).

The story seems to be quite different for the Clean Water Act. There also have been three attempts to estimate its aggregate benefits and costs, and again all three came to the same conclusion. However, in contrast with the Clean Air Act, the conclusion was that the costs of the water act exceeded its benefits. As one of the water act studies noted, "There are great uncertainties associated with both cost and benefit estimates that are not captured in the ... numerical results.... There are a number of tangible benefits for which monetary estimates have not been developed" (U.S. EPA 1994b, ES9-10). Nevertheless, the contrast with the studies of clean air costs and benefits is striking. Probably the major difference between the air and water acts in this context is that the benefits of the air act are primarily improved human health, whereas the benefits of cleaner water are primarily ecological and recreational. Society tends to place a higher value on health benefits.

A number of studies have taken a different approach to judging the efficiency of the pollution control regulatory system. Rather than looking at entire programs, they have looked at individual regulations. The results indicate significant inefficiencies in the system.

One of the most comprehensive of these studies examined abatement costs for the same pollutant across different industries (Hartman, Wheeler, and Singh 1994). A significant disparity in the marginal cost of abating pollution (that is, the cost of abating the next increment of pollution) among plants, firms, or industries indicates that maximum environmental protection is not being obtained for the amount of money being spent. The study examined air pollution abatement costs between 1979 and 1985 by economic sectors. The variances among sectors were often quite wide. For example, the marginal cost of abating a ton of lead was $46,612 in the food sector but only $11 in the nonmetal products sector; abating a ton of hydrocarbons cost the beverages industry $11,918 at the margin, but cost the pulp and paper industry only $20 (Hartman, Wheeler, and Singh 1994). The system could have reduced hydrocarbons much more efficiently by getting more reductions from pulp and paper plants and fewer from beverage makers. There are many such examples.

Several studies have looked at the cost-effectiveness of health regulations by measuring the dollars spent per life saved. These often are very partial studies because lives saved is usually based only on the results of risk assessments for cancer, thus ignoring other health benefits and all benefits other than health. Also, the numbers used in these studies are hugely uncertain. Nevertheless, the disparities among regulations are large enough to indicate that there is much room to improve cost-effectiveness. These studies also indicate that, to the extent that the goal of environmental regulation is to save lives, it often is more efficient to invest in nonenvironmental programs (such as encouraging motorists to wear seat belts) or in programs for early detection of serious diseases.

Finally, a number of studies have compared the costs of actual regulations with what the costs might have been for other environmental policy approaches, such as economic incentives. These comparisons suffer from all the problems of comparing an actual approach with a theoretical one. Nevertheless, they consistently find that approaches other than command-and-control regulation would be more efficient. A recent review of twenty-seven such comparisons found that command-and-control regulations cost one-and-a-half to more than twenty times as much as incentive-based approaches (Carlin 1992). The few examples of actual use of economic mechanisms tend to confirm the theoretical studies (see, for example, Burtraw 1995; U.S. GAO 1994). It is estimated that the sulfur dioxide emissions trading provisions of the 1990 Clean Air Act will save utilities between $0.5 billion and $2 billion per year by 2010 (Carlson and others 1997).

In sum, there is ample evidence that inefficiency is a problem for the pollution control regulatory system. The costs of some regulations may exceed their benefits by any definition of costs and benefits. While benefit-cost analysis is too crude a tool to use as the sole criterion of regulatory decisionmaking, when it indicates a large excess of costs over benefits a responsible decisionmaker needs to have a good reason for proceeding with the regulation. Efficiency is also a problem when approaches to compliance other than the current form of regulation may achieve the same goals at much lower cost.

SOCIAL VALUES

Some of the most vehement objections to environmental regulation have had nothing to do with its lack of efficiency or effectiveness. These objections have instead been based on the belief that regulations conflict with important social values—values such as privacy, the right to participate in decisions that affect one's life, due process, protection of private property, and nondiscrimination.

Environmental protection itself has become a basic social value. This is shown by numerous public opinion polls as well as by the public outcry that occurs whenever environmental values appear to be seriously threatened. The criticisms of pollution control efforts based on social values need to be placed in this context: the American people expect pollution to be controlled, and they expect their government to do it.

Evaluating programs based on their adherence to or violation of social values is difficult. There are no objective measures for most values, and judgments tend to be both strongly held and very subjective.

The idea that people should have a voice in decisions that affect them is probably the most salient social value in the regulatory system. The Administrative Procedures Act requires regulatory agencies to allow the public to comment on proposed rules and requires the agency to respond to comments. Many public participation mechanisms are employed by EPA, and both general laws (such as the Freedom of Information Act) and pollution control laws delineate methods for allowing public views to influence government decisions.

Although public participation is generally considered an unalloyed good, remarkably little serious analysis exists about what forms of participation work best, how the public can be most effective in its participation, and how participation mechanisms actually affect agency decisions. Our consideration of participation was handicapped by the lack of such analysis.

Environmental agencies at both the state and federal levels often have been in the forefront of experimenting with new forms of public participation. For example, EPA was one of the first federal agencies to use regulatory negotiation—a formal convening of interested parties—as a way of formulating regulations. More recently, the agency has turned to other forms of stakeholder participation in its Common Sense Initiative, XL projects, and place-based initiatives. Measuring the degree of participation, much less its effectiveness, is difficult, but our overall impression is that EPA is among the more "open" of all federal agencies.

EPA's openness does not seem to have earned it much trust or credibility. For the past twenty-five years, the American people have been increasingly mistrustful of all institutions, especially of their government. The corrosive effect on public actions is very clear with respect to environmental functions. People object to siting facilities and they ignore warnings of health threats; decisions about allowing or not allowing products to be sold are immensely difficult to implement because of people's mistrust of governmental authorities. Large portions of the public simply do not believe what government officials tell them. The problem is not unique to environmental officials, and there are no simple remedies.

Any long-run solution to the trust problem requires government officials to be completely open and honest in their dealings with the public. State environmental agencies vary widely in their degree of openness. The federal EPA, as we have noted, is relatively open, although it can be faulted for peddling a false certainty about some of its conclusions (for example, when giving only a single number as the conclusion of its risk assessments instead of a range). Congress often deceives the public by conveying the notion that environmental regulations can provide absolute safety: scientific findings indicate that for many, perhaps most, pollutants safety is a matter of degree and absolute safety is unobtainable. Both Congress and the executive branch shrink from conveying to the public the difficult trade-offs that environmental decisions often entail, and the public encourages such deception by wanting absolute safety at no economic cost in the same way it wants more government services and lower taxes. A better-educated public would reduce the temptation of government officials to practice such deception.

Protection from intrusiveness is another social value that the pollution control regulatory system is often accused of violating. We define intrusiveness as the characteristic of an action or requirement that results in an unwelcome or uninvited imposition on the public, including direct cost, lost time, restricted options, invasions of privacy, and inconvenience.

For this project, we developed a crude and quite subjective "intrusiveness index" to rank the major environmental programs. We found a fairly wide spectrum, with the National Environmental Policy Act being the least intrusive and air pollution control programs the most intrusive.

Like most social values, protection from intrusiveness is imbedded in a number of social values and cannot be considered in isolation. The general perception of recycling provides a good example. By our ranking, recycling programs are at least as intrusive as most other pollution control programs; yet they do not evoke the same negative reactions, perhaps because recycling is perceived as directly doing something good for the environment.

An opposite example is recordkeeping requirements, which many business people consider among the worst evils of the pollution control system. Probably this is because a lot of recordkeeping is viewed as redundant and unnecessary. The Paperwork Reduction Act requires most federal agencies to submit annually to the Office of Management and Budget (OMB) estimates of the time required by businesses and state and local governments to complete reports required by the agency (U.S. OMB 1989–1996). This information provides one measure of intrusiveness. In 1995, all federal agencies combined imposed a total of 6.9 billion hours of reporting. The Treasury Department imposed 5.3 billion of these hours, mostly through tax filing requirements imposed by the Internal Revenue Service. EPA imposed 104 million hours, more than the departments of Transportation and Education but less than several other federal agencies.

OMB does not report the paperwork requirements by individual regulatory program, but the Chemical Manufacturers Association analyzed the 1994 EPA data by program (CMA 1996, 5–14). It found a wide range. The water pollution control program imposed three times the burden of any other EPA program, requiring twenty-nine million hours of paperwork. Next were the air program with nine million hours and the Resource Conservation and Recovery Act program with seven million. The Safe Drinking Water Act program was lowest, imposing only eleven thousand hours of paperwork.

Environmental justice is a recent and quite sensitive social value affecting pollution control. Advocates for environmental justice have demanded increased regulatory oversight and more stringent environmental protection measures for minority and low-income communities and workers. Shortly after taking office, EPA Administrator Carol Browner listed environmental justice as one of her top priorities, and in 1994 President Clinton issued Executive Order 12898, Federal Actions to Address Environmental Justice in Minority Populations and Low-Income Populations.

EPA program offices have developed training programs for both EPA and state regulatory officials that examine how environmental justice concerns can be integrated into routine agency functions. In 1994 the agency began a small grants program to assist grassroots organizations and tribal governments in outreach work and in employing technical experts to analyze and interpret environmental data. However, it is not clear how much EPA can actually do to address perceived environmental inequities. Many of them spring from forces that are not within EPA's control—local land use practices, market dynamics, and state and local agency enforcement activities.

How environmental justice should be defined is not clear; neither is it clear, under most definitions, how serious the environmental justice problem is. Just the methodological problems of examining the issue are immense. For example, conflicting conclusions about equity emerge depending on the geographical unit used in an analysis. Measuring the distribution of environmental burdens by census tract can give quite different conclusions from measuring it by census block, zip code, or county (Glickman and Hersh 1995).

COMPARISON WITH OTHER COUNTRIES

For most of the twenty-five years that the environment has been on the agenda of all nations, the United States has been a leader. The United States was one of the first countries to launch a major national pollution control effort; it was a major force behind the 1972 Stockholm conference that put environment on the international agenda, and American environmental standards and institutions have been copied throughout the world.

Comparisons between the United States and other countries reveal a more mixed picture. A detailed comparison of pollution control standards is impossible because much of the impact of standards depends on technical details like measurement times and methods. Obviously, the degree of compliance with the standards also is critical. Most U.S. ambient and source discharge standards probably are at least as stringent as those in other countries, and the U.S. record of compliance is probably better. On some important specific requirements, such as removal of lead from gasoline, we know that the United States is ahead of most other nations.

Actual levels of environmental quality are difficult to compare for the same reasons that standards are hard to compare. The technical details of monitoring can make a large difference, and in most countries (including the United States) ambient monitoring data are sketchy. To the extent that data exist and can be roughly compared, and based on subjective impressions, ambient air and water quality in the United States compares favorably with that in most other countries.

The data on emissions of air pollutants tell a different story. The United States accounts for a huge proportion of the world's emissions of such major pollutants as carbon monoxide, nitrogen oxides, sulfur dioxides, and carbon dioxide (OECD 1996, 254–57).

Although some of the disproportionate air pollution emissions in the United States are attributable to the size of the U.S. population, geographic area, and economy, these factors do not explain a large part of the difference between the United States and other nations. When emissions data are controlled for such factors, a significant difference still exists between the United States and all other countries except Canada (OECD 1996, 254–55). Physical factors like climate may account for some of the difference, but it is hard to avoid the conclusion that the U.S. lifestyle is a significant part of the explanation. We live in bigger houses, drive more, and generally use more energy than other countries. While there are advantages to all this, it comes at an environmental cost.

The environmental impact of the U.S. lifestyle is not limited to air emissions. Per capita municipal waste generation in the United States is twice the level of such developed nations as Germany, Italy, and England (OECD 1996, 254–55).

The United States may be falling behind other countries with respect to environmental institutions. In September 1996, after several years of debate, the European Commission approved a directive that requires all countries in the European Union to adopt an integrated approach to pollution control, an approach that cuts across the lines of media (air, water, land) on which the U.S. system is based. In four or five years, when the European directive is fully implemented, the United States will be one of the few developed nations that has not adopted an integrated approach. It will be one of the few industrial countries issuing separate facility permits for air and water and not considering the tradeoffs among emissions to air, water, and land.

Market mechanisms also may be used more in other countries than in the United States, although in Europe pollution taxes often are used more to raise revenues than to affect the behavior of polluters. The United States has pioneered emissions trading and other important economic approaches to environmental problems, but other nations, especially in Europe, may be getting more environmental mileage from applying market mechanisms. Taxes on gasoline are a good example.

There are other innovative environmental programs from which the United States could learn some useful lessons. In addition to integrated approaches to pollution control in a number of countries, there are experiments with voluntary industry agreements on emissions reductions (the Dutch covenants), recycling requirements (the German law that makes packaging disposal the responsibility of the manufacturer), and energy conservation. Of course, major cultural differences need to be taken into account—in Japan, for example, saving face is a major motivating factor. Programs based on this incentive would not work well in the United States.

International trade is becoming increasingly important in the U.S. economy, and in recent years the numerous connections between trade and the environment have come to be recognized. To date, the U.S. pollution control regulatory system has probably not been a significant drag on American trade abroad, although U.S. firms clearly would benefit from a more efficient regulatory system (Jaffe and others 1995). Incorporating and reconciling environmental considerations into the international trade regime under the World Trade Organization is one of the major challenges facing the countries of the world.

ABILITY TO MEET FUTURE PROBLEMS

The problems addressed by the pollution control regulatory system are not static. Existing problems change for better or worse, and new problems are added to the agenda. In the background, a small but nagging chorus predicts that disaster is imminent and that most current programs resemble the proverbial rearranging of the deck chairs on the *Titanic*. The dynamic nature of environmental problems makes the ability to anticipate and deal with future problems an important criterion for evaluating the system.

One way of looking into the future is to look at the underlying trends that affect pollution levels. These trends do not all go in the same direction—some predictable trends will result in worse environmental quality, while others will improve the environment. The steady increase in vehicle miles traveled makes it harder to achieve satisfactory air quality. However, the trend towards "dematerialization," using less material to perform the same function (for instance, the transistor supplanting the vacuum tube), makes it easier to deal with waste problems. Individual problems must be examined in the light of individual trends—broad generalizations are likely to be misleading.

A look at energy use and other trends affecting air pollution emissions suggests that the level of most major pollutants will continue to decline over the next fifteen years (U.S. EPA 1994a; U.S. DOE 1995, 181). Two exceptions are carbon dioxide and particles. For both of these pollutants, emissions levels are closely related to energy consumption. EPA predicts that by 2010 emissions of particles (PM-10) will be almost 60 percent higher than they were in 1990 (U.S. EPA 1994a; U.S. DOE 1995, 181).

Water quality and water quantity are closely related because both affect water use. Because many people think that availability of water will be the dominant environmental problem of the 21st century, we looked at water supply projections for the United States. The picture is complicated by regional variations, lack of data, distinctions between withdrawal and consumption, instream versus non-instream uses, dry years versus wet years, and so forth. Several simplified conclusions are nevertheless possible. Projecting to 2040, only the Lower Colorado and Rio Grande river basins are likely to face major water shortages (Guldin 1989, 43). Shortages appearing elsewhere will be due more to institutional problems than to physical shortages. The overwhelmingly largest proportion of water use is for agricultural irrigation and, if mechanisms that facilitate transfer of water rights from agriculture to other uses can be encouraged, most of the shortages will disappear. This shift can occur without any significant damage to American agriculture. The most severe problem is in the Lower Colorado basin: even if all agricultural uses of water were stopped there, a water supply deficit would still exist (Guldin 1989, 43).

We foresee no startling changes or problems in the generation of municipal solid waste over the next couple of decades. Increases in population are likely to be offset by reduced waste disposal per person because of increased recycling. We also examined the problem of heavy metals, particularly chromium, mercury, and arsenic. Because heavy metals accumulate in the environment, they will, almost by definition, present an increasingly serious problem. We were frustrated in our attempts to estimate the rate of increase of metals in the environment because a large portion of the environmental load comes as a byproduct of other processes, especially the burning of coal. For example, 25 percent of total chromium emissions in the United States comes from fossil fuel combustion (Papp 1994, 37–38). This not only makes it difficult to estimate the magnitude of the problem but also indicates a possible weakness in the regulatory system: although a small amount of heavy metal emissions from fuel combustion is controlled by controlling particulate matter emissions, there are no direct controls on heavy metal emissions.

Completely new problems can appear on the environmental agenda. Nobody in 1970 thought that stratospheric ozone depletion was a problem. Five or ten years ago, not many people were worrying about endocrine disrupters. Identifying problems before they become major is inherently speculative, but it is important to try. EPA, to its credit, has in recent years recognized that future forecasting is an essential part of its responsibility. The purpose is not to predict the future but rather to examine the implications of current programs and to get a faster start on the next generation of problems.

Forecasting also highlights the degree to which pollution control is intertwined with policies in other areas such as energy, agriculture, and transportation. The U.S. government lacks ways to bridge across these different realms (with the notable exception of environmental impact statements). However, bridges need to be found, because the future of environmental quality will depend on what we do in these major economic sectors.

TOWARDS A BETTER SYSTEM

Given the scope and complexity of the pollution control system and the variety of criteria that we have used to evaluate it, we conclude not surprisingly that it has both strengths and weaknesses. Its greatest strength is its proven ability to reduce conventional pollutants generated by large point sources such as power plants and factories. It is a system that was developed to deal with the problems of the 1960s and 1970s, and it did a reasonably good job of addressing them.

In the course of dealing with those problems, the system has developed several other positive attributes. It opened the way to a variety of citizen efforts, ranging from recycling to regulatory negotiation and from citizen suits to monitoring the Toxics Release Inventory. Much remains to be learned about how to involve citizens effectively in government decisionmaking, but a number of the techniques that were first tried in the context of pollution control provide important lessons and models.

Congress, EPA, the states, and the private sector have developed other tools and approaches that help indicate the path of the future. Market-type mechanisms, particularly sulfur dioxide trading under the 1990 Clean Air Act Amendments, have proven to be as effective in practice as in theory, at least under the right circumstances.

EPA, in conjunction with the scientific community, also has made important contributions to the science necessary for dealing with environmental problems. It has advanced the art of conducting risk assessments and pioneered the approach of broad comparative risk assessment. The science of ecology, while still primitive and inadequately supported, has nevertheless benefited from federal and state efforts to understand and deal with pollution problems.

Despite these and other accomplishments, we conclude that the pollution control regulatory system has deep and fundamental flaws. There is a massive dearth of scientific knowledge and data. The system's priorities are wrong, it is ineffective in dealing with many current problems, and it is inefficient and excessively intrusive. Most of the participants in the system are aware of these defects.

There is no consensus about how to remedy these flaws. Not only do disagreements exist among the different interests concerned with pollution control, but even groups that seemingly have a common interest disagree with each other. There is no agreement among large corporations about decentralizing pollution control or about preserving the current regulations. There is no agreement among environmental groups about the utility of market mechanisms.

The United States does not need to wait for a consensus to act on these problems. Our political system is designed to negotiate agreements and find common ground. If we wait for consensus, we will wait forever.

Furthermore, some agreement exists about the principles that should guide changes in pollution control and about the characteristics of a pollution control system for the next century. The future system should be results-oriented, integrated, efficient, participatory, and information-rich.

Results-oriented. The current system is focused largely on how to control pollution rather than on whether pollution is actually being controlled. Technology-based standards have received the most criticism on this score, but ISO14000 (a fashionable industry nostrum that involves an international standard for corporate internal pollution control procedures) is similarly focused on means rather than ends. The system of the future must constantly ask whether human health and the natural environment are being adequately protected. Regulators need to set the standards, ensure that adequate data are available to know if the standards are being met, and take compliance measures if the standards are not being met. The means used to achieve the goals are secondary and should largely be left in the hands of the regulated parties.

Integrated. The fragmentation of the current system is a major factor in its lack of rational priorities, its inefficiency, and its difficulty in identifying and dealing with new problems. Within the next decade, most developed nations will have abandoned the medium-oriented system in favor of an integrated approach. The United States should not be saddled with an antiquated and cumbersome approach. An integrated approach, whether based on geographical area, economic sector, function (enforcement, research, and so forth), or some combination of these, is a prerequisite to most other basic reforms of the pollution control system.

Interagency or intersectoral integration is a different but equally important challenge. Future environmental quality will be determined by the nation's energy, agricultural, and transportation policies. Better ways to link environmental concerns and these other policy areas need to be instituted.

Efficient. The inefficiency of the current system should no longer be tolerated. Costs should be considered explicitly when establishing goals, and maximum flexibility should be allowed in achieving the goals. The use of market mechanisms should be a priority.

Participatory. Continuing efforts and experiments are required to encourage citizens to have some trust in their government and to participate in the decisionmaking process. However, public participation should not be used as an excuse for government to abandon its role as protector of the public interest.

Information-rich. The current system lacks all kinds of necessary information—scientific and economic information, information about actual environmental conditions (monitoring data), and information about whether programs are working (program evaluation). Recent events and trends—reduced spending for research and monitoring, the dismantling of the congressional Office of Technology Assessment—have made the situation worse. A new system has to recognize the need for information and provide the resources, incentives, and institutions to provide it.

The above characteristics implicitly assume that the federal government will continue to play a central role in controlling pollution and that that role will be based on congressionally enacted laws. Several considerations support a continuing federal role: pollution does not respect state or local borders, many of the most important problems require international action, lack of national standards would interfere with interstate commerce and might well result in a deterioration of environmental protection, and economies of scale exist for such functions as research. We think that law will continue to be centrally important because the kinds of incentives—especially financial gains or penalties—necessary to affect the behavior of potential polluters come primarily through law. Goodwill by itself is unlikely to move people to action. The question of what incentives are likely to change behavior is a necessary question to ask, and the answer frequently leads back to legislation.

The huge gap between agreeing to general characteristics and agreeing to specific changes and initiatives will not be easy to close, but the shortcomings of the existing system are so great that the nation needs to try. Agreement on general principles needs to be followed by the hard work of thinking through detailed policies and negotiating the political compromises necessary to enact and implement them. Failure to make the changes will be costly to the economy, to the environment, and to every citizen. The time to start considering these changes is now.

REFERENCES

Burtraw, Dallas. 1995. *Cost Savings Sans Allowance Trades? Evaluating the SO₂ Emission Trading Program to Date*. Discussion Paper 95-30-REV. Washington D.C.: Resources for the Future.

Carlin, Alan. 1992. *The United States Experience with Economic Incentives to Control Environmental Pollution*. Office of Policy, Planning, and Evaluation. Washington, D.C.: U.S. EPA.

Carlson, Curtis, Dallas Burtraw, Maureen Cropper, and Karen Palmer. 1997. *SO₂ Control by Electric Utilities: What Are the Gains from Trade?* Discussion Paper. Washington, D.C.: Resources for the Future.

CMA (Chemical Manufacturers Association). 1996. *Compliance Reporting Costs for EPA-Administered Environmental Laws: 1994*. Federal Relations Department, Economics Division.

Dodell, David, ed. 1994. *Blood Lead Levels: United States, 1988–1991*. MEDNEWS. Health Info-Com Network (HICNet) Medical Newsletter. http://ch.nus.sg/MEDNEWS/aug94/7361_6.html.

EESI (Environmental and Energy Study Institute). 1987. *The National Hazardous Waste Land Disposal Restrictions: Better Data, Clearer Policy Needed to Make it Work*. Washington, D.C.: Environmental and Energy Study Institute.

Glickman, Theodore, and Robert Hersh. 1995. *Evaluating Environmental Equity: The Impacts of Industrial Hazards on Selected Social Groups in Allegheny County, Pennsylvania*. Discussion Paper 95-13. Washington D.C.: Resources for the Future.

Guldin, R. W. 1989. *An Analysis of the Water Situation in the United States: 1989–2040*. Fort Collins, Colorado: USDA Forest Service.

Hartman, Raymond S., David Wheeler, and Manjula Singh. 1994. *The Cost of Air Pollution Abatement*. Environment, Infrastructure and Agriculture Division, Policy Research Department. Washington D.C.: World Bank.

Jaffe, Adam B., Steven R. Peterson, Paul R. Portney, and Robert N. Stavins. 1995. Environmental Regulation and the Competitiveness of U.S. Manufacturing: What Does the Evidence Tell Us? *Journal of Economic Literature* 33: 132–63.

NAPA (National Academy of Public Administration). 1995. *Setting Priorities, Getting Results, A New Direction for EPA*. Washington, D.C.: NAPA.

OECD (Organisation for Economic Co-Operation and Development). 1996. *Environmental Performance Review, United States*. Paris: OECD.

Papp, John F. 1994. *Chromium Life Cycle Study: U.S. Bureau of Mines Information Circular 9411*. Washington, D.C.: U.S. GPO.

Portney, Paul, ed. 1990. *Public Policies for Environmental Protection*. Washington, D.C.: Resources for the Future.

Reid, H. 1992. Testimony in Implementation of the Toxic Substances Control Act, Hearing before a Subcommittee on Toxic Substances, Environmental Oversight, Research and Development, U.S. Senate. March 25. Washington, D.C.: U.S. GPO.

Smith, Richard A., Richard B. Alexander, and Kenneth J. Lanfear. 1991. *Hydrology of Stream Water Quality: Stream Water Quality in the Coterminous United States: Status and Trends of Selected Indicators During the 1980s.* Water Supply Paper 2400. Washington, D.C.: U.S. Geological Survey.

Smith, Richard A., Richard B. Alexander and M. Gordon Wollman. 1987. *Analysis and Interpretation of Water Quality Trends in Major U.S. Rivers, 1974–1981.* Water Supply Paper 2307. Washington, D.C.: U.S. Geological Survey.

U.S. CEQ (Council on Environmental Quality). 1990. *Environmental Quality: The 21st Annual Report.* Washington, D.C.: U.S. GPO.

———. 1993. *Environmental Quality: The 24th Annual Report.* Washington, D.C.: U.S. GPO.

USDA. (U.S. Department of Agriculture). 1994. *Agricultural Resources and Environmental Indicators.* Agricultural Handbook No. 705. Natural Resources and Environment Division, Economic Research Service. Washington, D.C.: USDA.

U.S. DOE (Department of Energy). 1995. *Annual Energy Outlook 1995.* Energy Information Administration. Washington, D.C.: U.S. DOE.

U.S. DOI (Department of Interior). 1970–1990. *Metals and Minerals, Volume I. Minerals Yearbook.* Bureau of Mines. Washington, D.C.: U.S. GPO.

U.S. EPA (Environmental Protection Agency). 1987. *Unfinished Business: A Comparative Assessment of Environmental Problems.* February. Office of Policy Analysis and Office of Policy, Planning, and Evaluation. Washington, D.C.: U.S. EPA.

———. 1990. *Environmental Investments: The Cost of a Clean Environment.* Washington, D.C.: U.S. EPA.

———. 1991a. *Environmental Investments: The Cost of a Clean Environment, Report to the Administrator.* Washington, D.C.: U.S. EPA.

———. 1991b. *EPA's Pesticide Programs.* H-7506-C. Office of Pesticides and Toxic Substances. Washington, D.C.: U.S. EPA.

———. 1993. *State and Local Comparative Risk: Project Rankings Analysis.* Washington, D.C.: U.S. EPA.

———. 1994a. *National Air Quality and Emissions Trends Report, 1993.* EPA 454/R-94-026. Office of Air Quality Planning and Standards Research. Washington, D.C.: U.S. EPA.

———. 1994b. *President Clinton's Clean Water Initiative: Analysis of Benefits and Costs.* Office of Water. Washington, D.C.: U.S. EPA.

———. 1995a. *National Air Pollutant Emissions Trends, 1990–1994.* EPA-45/R-95-011. Office of Air Quality Standards. Washington, D.C.: U.S. EPA.

———. 1995b. *National Public Water System Supervision Program FY1993 Compliance Report.* EPA 812/R-94-001. Washington, D.C.: U.S. EPA.

———. 1995c. *National Water Quality Inventory, 1994 Report to Congress.* EPA 841-R-95-005. Office of Water. Washington, D.C.: U.S. EPA.

———. 1996a. *1994 Toxics Release Inventory: Public Data Release.* EPA 745-R-96-002. Office of Pollution Prevention and Toxics. Washington, D.C.: U.S. EPA.

———. 1996b. *Annual Report of the Office of Pollution Prevention and Toxics, FY1995.* EPA 745-R-96-005. Office of Pollution Prevention and Toxics. Washington, D.C.: U.S. EPA.

———. 1996c. *The Benefits and Costs of the Clean Air Act, 1970–1990, Draft.* Office of Air and Radiation. Washington, D.C.: U.S. EPA.

———. 1996d. *Characterization of Municipal Solid Waste in the United States: 1995 Update.* Office of Solid Waste and Emergency Response. USEPA 530-R-96-001. Washington, D.C.: U.S. EPA.

———. 1996e. Emergency Planning and Community Right-to-Know (EPCRA) Hotline. November. Personal communication with authors.

———. 1996f. *National Air Quality and Emissions Trends Report, 1995.* EPA 454/R-96-005. Office of Air Quality Planning and Standards. Washington, D.C.: U.S. EPA.

———. 1996g. Office of Radiation and Indoor Air. November. Personal communication with authors.

———. 1996h. Office of Water, Nonpoint Source Program. November. Personal communication with authors.

———. 1996i. Superfund Hotline, November. Personal communication with authors.

———. 1997. Office of Pollution Prevention and Toxics, New Chemicals Program, January. Personal communication with authors.

U.S. GAO (General Accounting Office). 1994. *Air Pollution: Allowance Trading Offers an Opportunity to Reduce Emissions at Less Cost.* GAO/RCED-95-30. Washington D.C.: U.S. GAO.

U.S. ITC (International Trade Commission). 1970–1990. *Synthetic Organic Chemicals: United States Production and Sales.* Washington, D.C.: U.S. ITC.

U.S. OMB (Office of Management and Budget). 1989–1996. *The Fiscal Year Information Collection Budget (ICB).* Office of Information and Regulatory Affairs, Total Information Collection Burden. Washington, D.C.: U.S. GPO.

———. 1996. *Budget of the United States Government Fiscal Year 1996.* Washington, D.C.: U.S. GPO.

Acronyms Used in This Book

AEC	Atomic Energy Commission (U.S.; now Nuclear Regulatory Commission)
APCA	Air Pollution Control Association
BOD	Biochemical oxygen demand
BATNEEC	Best available techniques not entailing excessive cost (U.K.)
BNA	Bureau of National Affairs (U.S.)
BTU	British thermal unit (heat measurement)
CAA	Clean Air Act (U.S.)
CAAA	Clean Air Act Amendments (U.S.)
CAP	Capacity assessment plan
CDC	Centers for Disease Control (U.S.)
CEC	Commission of the European Communities
CEQ	Council on Environmental Quality (U.S.)
CERCLA	Comprehensive Environmental Response, Compensation, and Liability Act (U.S.; Superfund)
CERCLIS	CERCLA Information System
CFCs	Chlorofluorocarbons, a class of ozone-depleting compounds
CMA	Chemical Manufacturers Association
CO	Carbon monoxide
CO_2	Carbon dioxide
CRA	Comparative risk assessment
CSO	Combined sewer overflow
CWA	Clean Water Act (U.S.)
CWS	Community water systems
DDT	A chlorinated insecticide, persistent in the environment
DM	Deutsche mark
DOD	U.S. Department of Defense
DOE	U.S. Department of Energy
DOI	U.S. Department of the Interior

DOL	U.S. Department of Labor
DOT	U.S. Department of Transportation
EC	European Community (EU as of 1994)
EEC	European Economic Community
EESI	Environmental and Energy Study Institute
EIS	Environmental impact statement
EPA	U.S. Environmental Protection Agency
EPCRA	Emergency Planning and Community Right to Know Act (U.S.)
EU	European Union (EC until 1994)
FDA	U.S. Food and Drug Administration
FIFRA	Federal Insecticide, Fungicide, and Rodenticide Act
FTC	Federal Trade Commission (U.S.)
FWS	U.S. Fish and Wildlife Service
GAO	General Accounting Office (U.S.)
GDP	Gross domestic product
GNP	Gross national product
HEW	U.S. Department of Health, Education, and Welfare (now HHS)
HHS	U.S. Department of Health and Human Services
HWCP	Hazardous Waste Cleanup Project
ITC	International Trade Commission (U.S.)
ITFM	Intergovernmental Task Force on Monitoring Water Quality
MCLs	Maximum containment levels
MSW	Municipal solid waste
NAAQS	National Ambient Air Quality Standards (U.S.)
NAPA	National Academy of Public Administration
NASQAN	National Stream Quality Accounting Network
NCSRP	National Contaminated Site Remediation Project (Canada)
NEPA	National Environmental Policy Act (U.S.)
NEPP	National Environmental Policy Plan (Netherlands)
NESHAP	National Emission Standard for Hazardous Air Pollutants, per CAA
NO_2	Nitrogen dioxide
NO_x	Nitrogen oxides
NOAA	National Oceanic and Atmospheric Administration (U.S.)
NPDES	National Pollutant Discharge Elimination System, per CWA
NPL	National Priorities List, per CERCLA
NPRM	Notice of Proposed Rulemaking (U.S.)
NRC	National Research Council (U.S.)
NRDC	Natural Resources Defense Council
NSTC	National Science and Technology Council (U.S.)

O_3	Ozone
OECD	Organisation for Economic Co-operation and Development
OMB	Office of Management and Budget (U.S.)
OSTP	Office of Science and Technology Policy (U.S.)
OSWER	Office of Solid Waste and Emergency Response (U.S. EPA)
OTA	Office of Technology Assessment (U.S.)
Pb	Lead
PCBs	Polychlorinated biphenyls; persistent environmental pollutant
PM-2.5	Particles (particulate matter) less than 2.5μ in diameter
PM-10	Particles less than 10μ in diameter
PSI	Pollutant standards index
RACT	Reasonably available control technology
RCRA	Resources Conservation and Recovery Act (U.S.)
RFF	Resources for the Future
RIAs	Regulatory impact analyses
SAB	Science Advisory Board (U.S. EPA)
SDWA	Safe Drinking Water Act (U.S.)
SEC	Securities and Exchange Commission (U.S.)
SIP	State implementation plan (U.S.)
SO_2	Sulfur dioxide
SO_x	Sulfur oxides
STARS	An internal EPA reporting and accounting system
TOE	Tons of oil equivalent (energy measure)
TRI	Toxics Release Inventory (U.S.)
TSCA	Toxic Substances Control Act (U.S.)
TSP	Total suspended particulates
UCC	United Church of Christ
U.K. DOE	Department of the Environment (U.K.)
UNEP	United Nations Environment Programme
USDA	U.S. Department of Agriculture
USGS	U.S. Geological Survey
VMT	Vehicle miles traveled
VOCs	Volatile organic compounds
WCED	World Commission on Environment and Development
WHO	World Health Organization
WRI	World Resources Institute
XL	"Excellence and Leadership" (a U.S. EPA reinvention project)

Index

Nuclear Regulatory Commission,
radiation control costs, 109
Nuclear waste management, 45, 107–9

Occupational Safety and Health
Administration, environmental
justice issues, 177
OECD countries, pollution control in.
See International pollution control
Office of Environmental Justice,
background of, 179
Office of Management and Budget
(OMB)
administrative decisionmaking, 28,
36, 37
costs of pollution control, 103
intrusiveness issues, 174, 175, 282
public involvement issues, 158
Office of Pesticide Programs,
administrative decisionmaking,
28
Office of Research and Development,
role of, 29
Office of Science and Technology
Policy (OSTP), administrative
decisionmaking, 36–37
Office of Solid Waste and Emergency
Response (OSWER), reporting
requirements, 35
Office of Technology Assessment
(OTA), dismantling of, 289
Office of Water, reducing pollution
levels, 69
Ohio
federal–state division of labor, 42
incentive-based pollution control
policies, 141
reducing levels of air pollution, 95,
275
OMB. *See* Office of Management and
Budget
Organisation for Economic
Co-operation and Development
countries, pollution control in. *See*
International pollution control
Organization of U.S. EPA. *See*
Personnel resources of U.S. EPA

OSTP. *See* Office of Science and
Technology Policy
OSWER. *See* Office of Solid Waste and
Emergency Response
OTA. *See* Office of Technology
Assessment
Our Common Future (WCED; 1987),
261–62
Our Stolen Future, 259
Oxygen concentrations in water,
reducing pollution levels, 70,
71–72, 73
Ozone (ground-level)
mortality rates and, 129
reducing levels of air pollution, 57,
59, 62, 64, 66, 67–68, 271
Ozone (stratospheric) depletion. *See
also* Air pollution
comparative risk assessments, 113,
114, 121, 277
environmental projections, 5, 286
fragmentation of pollution control
system and, 18
history of federal legislation, 14
international comparisons, 196

Packaging Ordinance of 1991
(Germany), 217
Paint, toxics in, 254, 256
Paperwork burden. *See* Reporting and
recordkeeping requirements;
Reporting systems
Paperwork Reduction Act
intrusiveness issues, 174, 282
public involvement issues, 169
Particulate matter (PM) emissions
comparative risk assessments, 115
environmental projections, 236–38,
285, 286
mortality rates and, 129
reducing pollution levels, 54, 57–58,
60, 62, 64, 66, 271
PCBs. *See* Polychlorinated biphenyls
Penalties, civil. *See* Litigation
(environmental)
Pennsylvania
environmental justice issues, 183